# Environmental Science and Engineering

**Series Editors**

Ulrich Förstner, Buchholz, Germany

Wim H. Rulkens, Department of Environmental Technology, Wageningen, The Netherlands

The ultimate goal of this series is to contribute to the protection of our environment, which calls for both profound research and the ongoing development of solutions and measurements by experts in the field. Accordingly, the series promotes not only a deeper understanding of environmental processes and the evaluation of management strategies, but also design and technology aimed at improving environmental quality. Books focusing on the former are published in the subseries Environmental Science, those focusing on the latter in the subseries Environmental Engineering.

Paulo Mendonça · Alberto T. Estevez · Yuan Chang
Editors

# Proceedings of 2023 International Conference on Green Building

Advances in Green Building

Springer

*Editors*
Paulo Mendonça
School of Architecture
Arts and Design (EAAD)
University of Minho
Guimarães, Portugal

Alberto T. Estevez
School of Architecture
International University of Catalonia
Barcelona, Spain

Yuan Chang
Central University of Finance
and Economics
Beijing, China

ISSN 1863-5520 ISSN 1863-5539 (electronic)
Environmental Science and Engineering
ISBN 978-3-031-43477-8 ISBN 978-3-031-43478-5 (eBook)
https://doi.org/10.1007/978-3-031-43478-5



This Springer imprint is published by the registered company Springer Nature Switzerland AG
The registered company address is: Gewerbestrasse 11, 6330 Cham, Switzerland

Paper in this product is recyclable.

# Conference Organization

## Conference Chairs

Paulo Mendonça, University of Minho (EAUM), Portugal
Alberto T. Estevez, Universitat Internacional de Catalunya, Spain
Yupeng Wang, Xi'an Jiaotong University, China

## Program Chairs

Yuan Chang, Central University of Finance and Economics, China
Patrick Tang, The university of Newcastle, Australia

## Program Co-chair

Mohammad Arif Kamal, Aligarh Muslim University, India

## Local Chair

Ju Liu, Malmö University, Sweden

## Technical Committee

Hedayat Omidvar, National Iranian Gas Company (NIGC), Iran
Hadi Erfani, The Society of Professional Engineers International UK, UK
Xiaodong Li, Tsinghua University, China
Bassam Qasim Abdulrahman, Darbandikhan Technical Institute, Iraq
Tao Wang, Chongqing University, China
Ahmed Abouaiana, La Sapienza University of Rome, Italy
Jingke Hong, Chongqing University, China
Didem Güneş Yılmaz, Bursa Technical University, Turkey
Yunfeng Chen, Purdue University, USA
Yingtao Qi, Xi'an Jiaotong University, China
Rute Eires, University of Minho, Portugal
Maria Manuela O. G. Almeida, University of Minho, Portugal
Lígia Silva, University of Minho, Portugal
Ayman N. Al-Dakhlallah (Tomah), Philadelphia University (Jordan), Jordan
Ali Karrech, School of Engineering, Australia
Kong Fah Tee, University of Greenwich, the United Kingdom
Paulo Mendonça, University of Minho (EAUM), Portugal
Mohammad Arif Kamal, Aligarh Muslim University, India
Ahmed Mancy Mosa, Al-Mansour University College, Baghdad
Ahmed Ajel Ali Al-Machtumi, Kufa University, Iraq
Hemaidi Zourgui Nadjib, École Nationale Supérieure des Travaux Publics, Algeria
Mady A. A. Mohamed, Effat University, KSA
Han-Yong Jeon, Inha University, Korea
Guang Ye, Delft University of Technology, Netherlands
Rudolf Hela, Brno University of Technology, Czech Republic
Zafar Said, University of Sharjah, United Arab Emirates
Hamed Badihi, Nanjing University of Aeronautics and Astronautics (NUAA), China
Hui Tong Chua, The University of Western Australia, Australia
Suhana Binti Koting, University of Malaya, Malaysia
Mo Kim Hung, University Malaya, Malaysia
Gordon Huang, University of Regina, Canada
Murat Eyvaz, Gebze Technical University, Turkey
Madhu Palati, BMS Institute of Technology and Management, India
António Gomes Martins, University of Coimbra, Portugal
Sumathy Krishnan, North Dakota State University, USA
Djamel Djenouri, University of The West of England (UWE), England
Ali Sarrafi Nik, Islamic Azad University, Iran
Qingpeng Man, Harbin Institute of Technology, China
Rufei Wang, Atelier Ten USA LLC, USA
Jingxiao Zhang, Chang'an University, China
Suryappa Jayappa Pawar, Motilal Nehru National Institute of Technology Allahabad, India

Xiaowei Luo, City University of Hong Kong, China
Andrea Pompigna, Università Telematica, Italy
Hongping Yuan, Guangzhou University, China
Yulong Li, Central University of Finance and Economics, China
Zezhou Wu, Shenzhen University, China
Masoud Taghavi, Technical and Vocational University (TVU), Iran
BEN AMMAR Ben Khadda, University of Biskra, Algeria

## Keynote Speakers

**Prof. Vivian W. Y. Tam**
Western Sydney University
**Speech Title:** Next Era in Recycling Concrete

**Prof. Kevin Zhang**
RMIT University
**Speech Title:** Circular Economy: The Transformation of Reclaimed Solid Waste into Construction Materials

**Prof. Enedir Ghisi**
Federal University of Santa Catarina, Brazil
**Speech Title:** Using Permeable Pavements to Harvest Stormwater for Non-Potable Water Uses in Buildings

**Prof. Thaddeus Chukwuemeka Ezeji**
Ohio State University, USA
**Speech Title:** Green Chemistry: A Case for Sustainable Production of Biofuels from Non-Food Substrates

# Preface

**Dear Readers,**

This book gathers them most relevant contributions to the 2023 International Conference on Green Building (ICoGB 2023), held in Malmö, Sweden on May 19–21, 2023

The building sector has an important role to play in moving toward a more sustainable future by reducing emissions, waste, and environmental impacts. In this new century, the world faces unprecedented challenges related to climate change, resource scarcity, and environmental degradation. The goal of the ICoGB 2023 was to serve as a catalyst for real progress in green building at both conceptual and practical levels.

Researchers and scholars from across the globe have unveiled their groundbreaking findings, such as our keynote speakers, Prof. Vivian W. Y. Tam from Western Sydney University and Prof. Kevin Zhang from RMIT University. They candidly shared their invaluable experiences and offered profound insights into the latest trends of green building. By fostering international connections and collaborations, the conference explored the innovation needed for promoting a greener built environment in all the continents.

I would like to express my sincere gratitude to the program chairs for their invaluable guidance, which has ensured high-quality submissions, and to the technical committee members for their diligent and accurate peer reviews.

Furthermore, I extend my deepest appreciation to the organizing committee which, with their professional devotion and excellence, was able to assure that all participants of ICoGB 2023 experienced an enriching and fruitful conference. Finally, I have to acknowledge all the active participants, onsite and online, who were able to promote the discussion between presenters and audience, enriching the scientific quality of the event with their valuable comments and questions.

Guimarães, Portugal

Paulo Mendonça
ICoGB 2023 Conference Chair

# Contents

# Chapter 1
# A Review on Advancements of Novel Building Thermal Insulation Materials: Zirconia Aerogels

**Zhang Liu, Xiao-Juan Li, Qian Ma, Li Cheng, He Qi, Yuan Yuan, Zhonghua Zhang, and Tao Feng**

**Abstract** Zirconia ($ZrO_2$) aerogels have three-dimensional, cross-linked network structures where the main component is $ZrO_2$. They have characters of acid–base duality, redox properties, high specific surface areas, high porosities, and high levels of thermal and chemical stability. Due to the above-mentioned characters, $ZrO_2$ aerogels can be widely used in the field of construction as fireproof and thermal insulation, and has great potential in the field of building energy conservation. This paper aims to provide a timely review on recent advancements in the preparation, modification, and applications of $ZrO_2$ aerogels. The research and applications of the direct sol–gel and organic acid-assisted sol–gel methods remain under-developed, however, these two methods have the greatest potential in future preparations of aerogels because

Z. Liu · L. Cheng
Institute of Future Human Habitats, Tsinghua Shenzhen International Graduate School, Tsinghua University, Shenzhen, PR China
e-mail: izliu@outlook.com

L. Cheng
e-mail: cjackl@126.com

Z. Liu · X.-J. Li · Q. Ma · L. Cheng · H. Qi · Y. Yuan · Z. Zhang (✉) · T. Feng
China Construction Science and Technology Cooperation, Shenzhen, China
e-mail: zh_zhang_cscec@outlook.com

X.-J. Li
e-mail: gracelxjsx@163.com

Q. Ma
e-mail: serenama1993@163.com

H. Qi
e-mail: qihe@cscec.com

T. Feng
e-mail: t_feng_cscec@outlook.com

Z. Zhang
Tsinghua University, Beijing, China



P. Mendonça et al. (eds.), *Proceedings of 2023 International Conference on Green Building*, Environmental Science and Engineering,
https://doi.org/10.1007/978-3-031-43478-5_1

of their low toxicity and low safety risk. Atmospheric drying is the most active field of research because of its low cost and compatibility with conventional atmospheric pressures. The widely used modification methods include (1) doping modification and (2) adding a binary composite, and the latter is comparatively more common because $ZrO_2$ aerogels prepared via this method exhibit improved stability at high temperatures compared with corresponding aerogels prepared by doping.

**Keywords** Zirconia aerogel · Sol–gel process · Supercritical drying · Aerogel thermal insulation · Aerogel catalyst · Aerogel adsorbent

## 1.1 Introduction

Aerogels are lightweight and nanoporous, and one can obtain them from the corresponding wet gels by replacing the liquid in the gels with air (Du et al. 2013). The aerogel structure is an open, cross-linked network with 80% to 99.8% porosity (less than 50 nm). Accordingly, aerogels exhibit an unusual combination of properties for one material, such as low density (0.004–0.500 g/cm$^3$), low thermal conductivity (0.017–0.05 W m$^{-1}$ K$^{-1}$), and high specific surface area (500–1200 m$^2$/g) (Bangi et al. 2019). Kistler fabricated the first generation of aerogels (including those based on $SiO_2$, $Al_2O_3$, or $SnO_2$) in 1931 (Kistler 1931). After 90 y of development, the types, chemical composition, and microstructure units of aerogels have become increasingly diverse. One can classify current aerogels as those based on metal oxides (Sui and Charpentier 2012), organics (Tao et al. 2008), carbon (Hu et al. 2019), carbide (Yu et al. 2018), and composites formed by fibers and/or whiskers (Guan et al. 2017; Pierre and Pajonk 2002).

$ZrO_2$ aerogels exhibit the properties of zirconia and conventional aerogels. Figure 1.1 shows a typical appearance and microstructure. $ZrO_2$ has the following: high melting point, acid–base duality, low density, tailorable nanostructure, low thermal conductivity, redox properties, high specific surface area, high porosity, and high level of chemical stability (Srikanth and Madhu 2017). Accordingly, researchers widely use $ZrO_2$ for thermal insulation, catalysts and catalyst carriers, dye-sensitized solar cell electrodes, solid oxide fuel cells, and purifiers in air and water (Zhu et al. 2015). $ZrO_2$ prepared from zirconium isopropoxide through the sol–gel process and supercritical drying in a benzene or isopropanol solution (as reported by Teichner et al. (1976)) is an active area of research and a focus of aerogel-pertinent studies (Teichner et al. 1976; Hurwitz et al. 2020; Shen 2017).

This paper reviews and discusses recent syntheses, modifications, and applications of $ZrO_2$ aerogels (such as thermal insulation, catalysis, and adsorption). We also propose potential applications of and prospects for $ZrO_2$ aerogels.

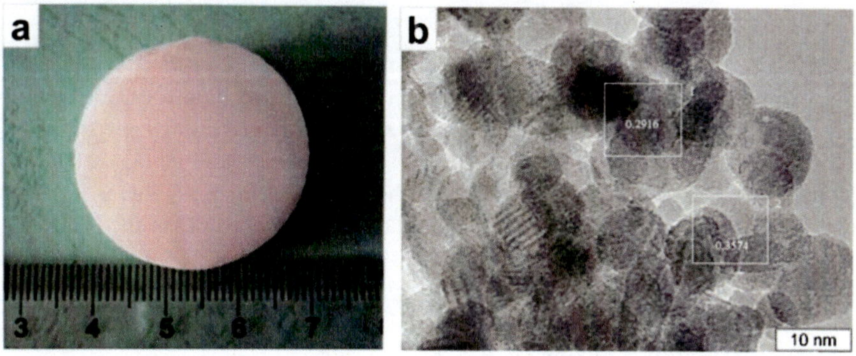


**Fig. 1.1** **a** Sample of $ZrO_2$ (Li et al. 2013) and **b** transmission electron microscopy image of $ZrO_2$ (Veselovskaya et al. 2021)

## 1.2 Preparation of $ZrO_2$ Aerogels

One can obtain $ZrO_2$ aerogels from the corresponding zirconium-containing precursor after continuous hydrolysis and polycondensation. The preparation (Fig. 1.2) consists of three steps: (1) preparing the wet gel, (2) aging, and (3) drying. In the first step, one dissolves the precursors (such as inorganic salts or metal alkoxides) in water or organic solvents. Then one adds various chemical additives and coagulants to hydrolyze the precursors in the reaction solution to form primary particles with a size of ca. 1 nm. Next, polycondensation continues to form the sol. In the second step, one ages the prepared wet gel in the mother liquor. This step is a continuation of the hydrolysis and polycondensation, strengthens the network structure of the gel, and prevents excessive shrinkage and cracking during the subsequent drying. Common aging solutions are mixed liquids of alcohol and water, and the alcohol solution of the precursor (Nadargi et al. 2009). The third step is to discharge the liquid in the gel network through e.g. supercritical drying, vacuum freeze-drying, or atmospheric drying. Afterward one obtains an aerogel that resists deformation.

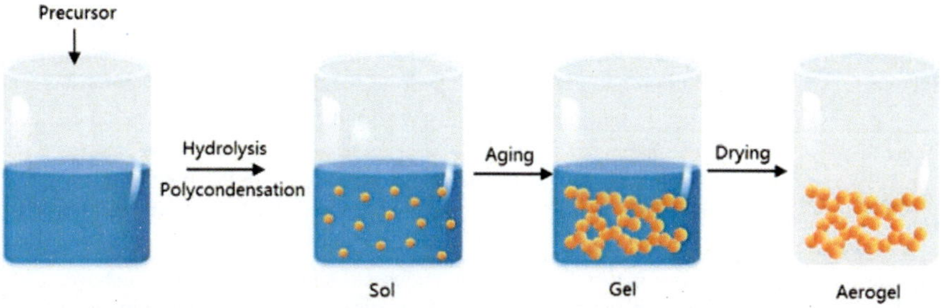


**Fig. 1.2** Aerogel synthesis

## 1.2.1 *Preparation of $ZrO_2$ Wet Gels*

The methods for preparing $ZrO_2$ wet gels mainly include zirconium alkoxide hydrolysis, electrolysis, precipitation, alcohol–water solution heating, propylene oxide (PO) addition, inorganic dispersion sol–gel, direct sol–gel, and organic acid-assisted sol–gel methods. Brief descriptions of each method are presented in Table 1.1. These methods, and pertinent advantages and limitations, are described in the following subsections.

### 1.2.1.1 Zirconium Alkoxide Hydrolysis

This method generally uses organic zirconium alkoxide as the raw material. One can prepare the wet gel by adding a sufficient quantity of acid catalyst to hydrolyze and

**Table 1.1** Method for Preparation of $ZrO_2$ Wet Gels

| Method | Raw material | Preparation process |
|---|---|---|
| Zirconium Alkoxide Hydrolysis | Organic zirconium alkoxide | Adding a acid catalyst to hydrolyze and polymerize the alkoxide, and removing the solvent from the wet gel to get $ZrO_2$ aerogel |
| Electrolysis | Aqueous solution of a metal chloride | Electrolyzing an aqueous solution of a metal chloride followed by drying |
| Precipitation | Ammonia water and inorganic zirconium salt solution | Adjusting the pH of the inorganic zirconium salt solution to alkaline for precipitating colloidal $Zr(OH)_4$, then generating $ZrO_2$ aerogel through sol–gel and drying processes |
| Alcohol–Aqueous Solution Heating | Inorganic zirconium salt/zirconyl nitrate | Heating inorganic zirconium salt/zirconyl nitrate in an ethanol/aqueous solution to form a colloid, then exchanging the colloid with an alcohol to obtain an alcohol gel, and drying the alcohol gel by using a supercritical fluid |
| PO Addition | Zirconium n-propoxide (and/or zirconium n-butoxide) | The sol–gel process is obtained in a supercritical $CO_2$ environment |
| Inorganic Dispersion Sol–Gel | Inorganic metal solutions and PO | The sol–gel process is obtained from reaction of the hydrous metal ions with the decomposing hydroxide radicals |
| Direct Sol–Gel | Inorganic metal solutions and polyacrylic acid (PAA) | The process is similar to the process above, however, compared with PO addition, PAA could enhance the crosslinking between the colloidal particles |
| Organic Acid-Assisted Sol–Gel | Inorganic metal solutions and organic acids | The process is similar to the process above, however, compared with PO addition, organic acids has lower toxicity and low safety risk to preparing bulk $ZrO_2$ aerogels |

polymerize the alkoxide. Subsequent removal of the solvent (in the pores) from the wet gel results in a $ZrO_2$ aerogel.

Bedilo et al. used ethanol-soluble zirconium n-butoxide as a precursor and added an aqueous alcohol containing nitric acid as a catalyst. After stirring, aging at room temperature, and high-temperature supercritical drying, they obtained a $ZrO_2$ aerogel with a specific surface area of 385 $m^2/g$ (Bedilo and Klabunde 1997). They studied the sensitivity of various parameters (such as the concentration of zirconium n-butoxide and nitric acid, water content, aging time, supercritical drying, and calcination temperature) on the specific surface area and pore volume of the resulting gel (Bedilo and Klabunde 1997).

Torres-Rodríguez et al. prepared a $ZrO_2$ wet gel with the following procedure: (1) used zirconium n-propoxide as the zirconium source; (2) added an n-propanol/water solution to the mixed solution containing nitric acid, zirconium n-propoxide, and n-propanol; (3) stirred vigorously for 5 min; and (4) left the reaction mixture at room temperature for 1 d. Accordingly, they proposed a comprehensive reaction mechanism for the formation of their $ZrO_2$ wet gel network (Torres-Rodríguez et al. 2019).

As a classic method for preparing $ZrO_2$ wet gel, zirconium alkoxide hydrolysis has the advantages of producing a product of high specific surface area, high purity, and uniform particle size dispersion; but the precursor is expensive, flammable, and toxic. Furthermore, it is difficult to control the hydrolysis of organic zirconium alkoxide, which further limits applications of the method (Suh and Park 1996). As a result, replacing zirconium alkoxide with inorganic zirconium salts as precursors is increasingly common.

#### 1.2.1.2 Electrolysis

This method prepares $ZrO_2$ wet gels by electrolyzing an aqueous solution of a metal chloride followed by drying. The reaction is mild, environmentally friendly, and suitable for large-scale preparation of aerogels with a high specific surface area.

Zhao et al. (2007) used $ZrOCl_2$ as a raw material to prepare a $ZrO_2$ wet gel by electrolysis combined with a sol–gel method. They synthesized a $ZrO_2$ aerogel with a high specific surface area through supercritical $CO_2$ drying (and/or freeze-drying). During electrolysis, Zr4+ coordinated with the OH− in the solution to generate $[Zr_3(OH)_4]^{8+}$, $[Zr_3(OH)6Cl_3]3+$, and [Zr4(OH)8]8+ as metal cluster complexes, which further condensed to form inorganic polymers. When the concentration of the polymer reached a critical value, the sol particles formed a stable gel through hydroxyl bridges and oxygen bridges. The aerogel prepared by supercritical drying was a transparent block with an average pore diameter of 9.7 nm and a specific surface area of ca. 640 $m^2/g$, whereas the aerogel prepared by freeze-drying was an opaque white powder with an average pore diameter of 0.59 nm and a specific surface area of 400 $m^2/g$.

#### 1.2.1.3 Precipitation

This method uses ammonia water to adjust the pH of the inorganic zirconium salt solution to alkaline for precipitating colloidal $Zr(OH)_4$. Then one generates a $ZrO_2$ aerogel through sol–gel and drying processes (Huang et al. 1998).

Liang et al. used zirconyl nitrate as a raw material and added ammonia to adjust the pH to 9.6 to form a colloidal precipitate of $Zr(OH)_4$ (Liang et al. 2005). Then they obtained an alcohol gel after aging and suction filtration. One can also obtain a $ZrO_2$ aerogel powder after supercritical $CO_2$ drying and calcination at 110 °C for 2 h. The specific surface area of this powder was 279.6 $m^2/g$ and the pore volume was 0.8 $cm^3/g$ (Liang et al. 2005).

At present, preparation of $ZrO_2$ aerogels by precipitation requires relatively cheap raw materials, but the synthesis is usually too fast, which leads to agglomeration and imparts difficulties to obtaining an aerogel with a uniform network structure.

#### 1.2.1.4 Alcohol–Aqueous Solution Heating

This method uses inorganic zirconium salt/zirconyl nitrate as raw materials, which one heats in an ethanol/aqueous solution to form a colloid. Exchanging the colloid with an alcohol to obtain an alcohol gel, and drying the alcohol gel by using a supercritical fluid, results in a $ZrO_2$ aerogel (Hu et al. 2001; Sun et al. 1997).

Bai et al. prepared $ZrO_2$ aerogels with a specific surface area of 668.2 $m^2/g$ by using $ZrO(NO_3)_2{\cdot}2H_2O$ as the raw material and utilizing alcohol–aqueous solution heating combined with supercritical fluid drying (Bai et al. 2006). They hypothesized that the preparation conditions (such as the alcohol/water ratio, zirconium salt concentration, and aging time) considerably impact the specific surface area and pore structure parameters of $ZrO_2$ aerogels.

Zhong et al. (2014) used $ZrO(NO_3)_2{\cdot}2H_2O$ as the zirconium source and urea as the gel accelerator. They obtained $ZrO_2$ aerogels with a specific surface area of 445 $m^2/g$ by alcohol–water solution heating, in conjunction with supercritical ethanol drying.

#### 1.2.1.5 PO Addition

This method completes the sol–gel process in a supercritical $CO_2$ environment and has the advantages of a high conversion rate, being environmentally friendly, and ease of scale-up.

Sui et al. produced porous $ZrO_2$ aerogels with nanostructures by using zirconium n-propoxide (and/or zirconium n-butoxide) as raw materials and acetic acid as a catalyst, in conjunction with the direct sol–gel method in supercritical $CO_2$ (Sui et al. 2006). The specific surface area of the aerogel reached 399 $m^2/g$. The coordination of acetate and metal ions can reduce the hydrolysis rate of zirconium alkoxide in water, thereby slowing down sol–gel production and facilitating formation of a uniform nanostructure.

#### 1.2.1.6 Inorganic Dispersion Sol–Gel

In 2001, Gash et al. (2001a, 2001b) prepared a series of trivalent/tetravalent metal oxide (such as Fe3+, Al3+, In3+, Ga3+, Zr4+, and Hf4+). They used inorganic metal solutions as precursors and PO as a gelation inducer. The added PO underwent a slow ring-opening reaction with H+ to remove free protons from the system, thereby facilitating reaction of the hydrous metal ions with the decomposing hydroxide radicals. An advantage of this method is that researchers can thereby modulate the condensation rate. The slow progress of the gelation facilitates formation of an aerogel with a uniform, three-dimensional network structure (Yu et al. 2014).

Guo et al. (2011a) prepared a lightweight $ZrO_2$ aerogel by using zirconyl nitrate as a precursor, PO as a gel accelerator, and ethanol/water as a mixed solvent, in conjunction with atmospheric drying. The resulting aerogel had an apparent density of 202.08 kg/m3 and a specific surface area of 645 $m^2/g$. Accordingly: (1) PO facilitates gel formation, and (2) one can modulate the reaction and gel state by changing the quantity of PO added.

Li et al. (2013) discussed the influence of the quantity of PO added on the microstructure of bulk $ZrO_2$ aerogels. As the molar ratio of PO to Zr increased from 0 to 2.5, the gel time decreased and the connectivity of the $ZrO_2$ aerogel network increased. The average pore radius increased from 6.6 to 15.1 nm, and then decreased to 8.5 nm. The pore volume increased from 1.2 to 1.87 $cm^3/g$, and then decreased to 1.44 $cm^3/g$.

Liu et al. (2019a) prepared polyzirconium acetylacetonate (PAZ)– $ZrO_2$ aerogel by using PAZ as the precursor, adding formamide, and adding PO as inducer. They compared the physical properties of PAZ–$ZrO_2$ aerogels with those of zirconium oxychloride (ZOC)–$ZrO_2$ and zirconyl nitrate (ZON)–$ZrO_2$ aerogels. Compared with ZOC–$ZrO_2$ and ZON–$ZrO_2$ aerogels, PAZ–$ZrO_2$ aerogels exhibited higher mechanical strength (0.12–0.23 MPa) and lower density ($0.12 \pm 0.005$ g/$cm^3$). After heating at 1000 °C for 2 h, the specific surface areas of PAZ–$ZrO_2$, ZOC–$ZrO_2$, and ZON–$ZrO_2$ aerogels were 236, 210, and 223 $m^2/g$, respectively.

Preparing $ZrO_2$ aerogels by adding PO does not make use of organic zirconium alkoxide as a precursor, and in so doing reduces the cost and avoids water sensitivity as well as hydrolysis of zirconium alkoxide. Furthermore, the catalysis requires no acids or alkalis. The method also expands the scope of preparing bulk aerogels.

#### 1.2.1.7 Direct Sol–Gel

This method used a small quantity of polyacrylic acid (PAA) as a dispersant and gel inducer to improve the performance of the resulting bulk aerogel. Compared with PO addition, one enhances the crosslinking between the colloidal particles (Xiao et al.

2008; Du et al. 2009). During gelation, PAA (with active carboxyl groups) plays a bridging role by facilitating nucleation, growth, and adhesion of the polymer chains.

Mu (2017) prepared a $ZrO_2$ aerogel by using zirconium tetrachloride as the precursor and PAA as a modifier combined with the inorganic dispersion sol–gel method. Mu investigated the effect of the quantity of PAA on the $ZrO_2$ aerogel performance. When the molar ratio of zirconium salt/PAA was 24:1, the density of the $ZrO_2$ aerogel was 0.2101 g/cm$^3$ (minimum value). After calcination at 900 °C, the specific surface area was 68.5 m$^2$/g (maximum value).

Currently, research on preparing aerogels by the inorganic dispersion sol–gel method remains under-developed. More studies are required on its mechanism and applications.

#### 1.2.1.8 Organic Acid-Assisted Sol–Gel

This a comparatively new method that uses organic acids (such as citric acid and aspartic acid) instead of PO as a gel accelerator, which imparts advantages of low toxicity and low safety risk to preparing bulk $ZrO_2$ aerogels (Zhang et al. 2015).

Wang et al. (2018) used $ZrOCl_2 \cdot 8H_2O$ as the precursor zirconium source and an organic acid (succinic acid, malic acid, aspartic acid, or mercaptosuccinic acid) as a gel accelerator to prepare bulk $ZrO_2$ aerogels. They investigated the gelling mechanism; one can combine the functional groups [such as hydroxyl (sulfhydryl) and carboxyl] in the organic acids with Zr4+ in the form of covalent bonds or coordination bonds, and thus form a gel with a stable, rigid network structure. The specific surface area of the resulting $ZrO_2$ aerogel prepared by using the aspartic acid gel accelerator reached 330 m$^2$/g.

### *1.2.2 Drying Methods of $ZrO_2$ Aerogels*

Drying is critical to preparing $ZrO_2$ aerogels. At present, the main drying technologies for $ZrO_2$ wet gels are supercritical drying, freeze-drying, and atmospheric drying.

#### 1.2.2.1 Supercritical Drying

Supercritical drying was initially the most commonly used drying technology. One gradually removes the liquid in the wet gel through a reaction-like dissolution and replacement in a supercritical state (with supercritical fluid). Because there is no longer a gas–liquid interface under critical conditions, the capillary force in the microporous structure of the wet gel is negligible. As a result, supercritical drying can prevent shrinkage and collapse of the aerogel during drying.

The commonly used drying media in supercritical drying are $CO_2$ (Schäfer et al. 2013), ethanol (Wu et al. 2004; Cao et al. 2002), and n-propanol (Bianchi et al. 1993;

Lecomte et al. 1998). Because the critical temperatures of ethanol and n-propanol are high (243 °C and 263.5 °C, respectively), it is dangerous to use ethanol and/or n-propanol in supercritical operations. Alternatively, the critical temperature of $CO_2$ is 31 °C. Furthermore, supercritical drying with $CO_2$ has advantages of being environmentally friendly, nontoxic, cheap and straightforward to obtain, and recyclable. Therefore, liquid $CO_2$ is now the most popular supercritical drying medium researchers use to prepare $ZrO_2$ aerogels.

#### 1.2.2.2 Freeze-Drying

This method dries the aerogel by freezing the $ZrO_2$ gel and then subliming the solvent at low temperatures and pressures. During freeze-drying, although there is no surface tension at the gas–liquid interface, the liquid–solid phase transition is accompanied by volume changes and the formation of crystal grains, which inevitably corresponds to local destruction of the gel network structure. One obtains the aerogel in powder form. Accordingly, even though freeze-drying is straightforward, the characteristics of the resulting $ZrO_2$ aerogel are poor compared with supercritical drying, in terms of the specific surface area and porosity (Zhao et al. 2007).

Presently, there are few studies in the literature on $ZrO_2$ aerogels (with stable structures and properties) that researchers prepared by freeze-drying. One can improve the toughness of aerogel materials by constructing organic and inorganic networks, and in so doing expand the scope of freeze-drying. Le (Le et al. 2018) prepared cellulose/$ZrO_2$ composite aerogels with various $ZrO_2$ contents by using a synthetic cellulose aerogel as a matrix and $ZrOCl_2$ as a precursor, in conjunction with a sol–gel method and freeze-drying. Accordingly, the maximum specific surface area of the aerogel reached 154 $m^2/g$.

#### 1.2.2.3 Atmospheric Drying

The high energy consumption and cost of traditional supercritical drying technologies impedes large-scale production of $ZrO_2$ aerogels. Accordingly, low-cost and mild atmospheric drying is an active area of research (Bangi et al. 2019). The key to preparing $ZrO_2$ aerogels through atmospheric drying is to avoid the shrinkage and cracking of the gel caused by capillary forces during drying.

One can group the commonly used atmospheric drying measures as follows: (1) modification of the surface of the gel by using a silane coupling agent to reduce the shrinkage caused by the hydroxyl polycondensation of adjacent surfaces in the gel skeleton during drying; (2) utilization of a low-surface-tension solvent to replace the high-surface-tension solvent in the gel pores; (3) utilization of surfactants to improve the size and uniformity of the pores in the gel, and to reduce the shrinkage and rupture caused by uneven stress (Bangi and Park 2018); and (4) utilization of a secondary gel to build a more-stable pore structure and minimize the collapse of the pores in the aerogel (Li et al. 2016).

By combining PO addition and alcohol–water solution heating, Guo et al. (2011b) prepared $ZrO_2$ aerogels at conventional atmospheric pressures. They used zirconium inorganic salt as the precursor, added chelating agents and drying-control chemical additives, and treated the wet gel with a silane coupling agent.

Bangi et al. (2020) used zirconium n-propoxide as the precursor and dissolved it in a mixed solution of n-propanol and sulfuric acid, then added deionized water to the solution. The researchers then replaced n-propanol and water with ethanol or n-hexane in the gel pores. Finally, they obtained $ZrO_2$ alcogels after the reaction between the gel surface and hexamethyldisilane. The density and specific surface area of the resulting $ZrO_2$ aerogel were 0.54 g/cm$^{-3}$ and 328 m$^2$/g, respectively.

Jung et al. (2017) used (1) zirconium n-propoxide as the precursor; (2) n-propanol as the solvent; (3) acetic acid as the chelating agent; (4) n-hexane as the low-surface-tension solvent; (5) hexamethyldisilazane as the silicon-based surface modifier; and (6) the block copolymer Brij S10, triblock copolymer Pluronic P-123, and cetyltrimethylammonium bromide as surfactants to prepare $ZrO_2$ aerogels; in conjunction with drying at atmospheric pressure. They investigated the role the surfactants played in the formation of the pore structure. The aforementioned three surfactants had positive effects as follows: (1) prevented collapse of the pore structure, (2) increased the specific surface area, and (3) improved the high-temperature stability of the $ZrO_2$ aerogel. Among the surfactants, cetyltrimethylammonium bromide exhibited the best performance in terms of preventing collapse of the pore structure.

## 1.3 Modification Methods for $ZrO_2$ Aerogels

There are three crystal forms of pure $ZrO_2$ at conventional atmospheric pressures: (1) monoclinic phase (m-$ZrO_2$), (2) tetragonal phase (t-$ZrO_2$), and (3) cubic phase (c-$ZrO_2$). Specifically: (1) m-$ZrO_2$ is the stable low-temperature phase; (2) t-$ZrO_2$ gradually forms when the temperature is higher than 1170 °C; and (3) c-$ZrO_2$ forms when the temperature is higher than 2370 °C (Garvie 1978; Khajavi et al. 2014). The transformation of the aforementioned crystal forms is reversible.

As the temperature increases, the crystal structure of $ZrO_2$ changes from a monoclinic phase to a tetragonal phase, which is accompanied by volume shrinkage of 7–9%, corresponding to destruction of the pore structure of the $ZrO_2$ aerogel and thereby affecting the corresponding performance (Wang 2014). Therefore, many researchers have studied doping additives and binary compounds to improve the stability of $ZrO_2$ aerogels at high temperatures.

### 1.3.1 Doping Modifications

This method is to dope $ZrO_2$ with divalent or trivalent cations (M3+ or $m^2$ + ). The ionic radius of the divalent or trivalent cations is similar to that of zirconium ion, which can replace Zr4+ in the sublattice and stabilize the structure of $ZrO_2$ by generating oxygen vacancies. A representative divalent or trivalent cation is Y3+ (Hurwitz et al. 2020).

Chervin et al. (2006) prepared $Y_2O_3$-stabilized $ZrO_2$ (YSZ) aerogels by using the sol–gel method and supercritical $CO_2$ drying, by utilizing $ZrCl_4$ and $YCl_3{\cdot}6H_2O$ as raw materials and PO as a gelling agent. The specific surface area of the resulting aerogel was 453 $m^2$/g. The specific surface area of the YSZ aerogel was greater than 100 $m^2$/g after calcination at 550 °C when the proportion of added Y2O3 was 3–15 mol%.

Chao et al. (2016) used $ZrOCl_2$·8H2O and $Y(NO_3)_3{\cdot}6H_2O$ as raw materials (in conjunction with the sol–gel method and supercritical drying) to prepare crack-free, massive YSZ aerogels. They studied the change in the crystal form of the aerogel at various sintering temperatures and quantities of added Y2O3. Doping with $Y_2O_3$ inhibited the transformation from t-$ZrO_2$ to m-$ZrO_2$, and the growth of the aerogel grains was inhibited with increasing $Y_2O_3$ content. When the proportion of $Y_2O_3$ reached 8 wt.%, t-$ZrO_2$ remained stable at 1200 °C.

Torres-Rodríguez et al. (2019) used $Zr(OC_3H_7)_4$ and $Y(NO_3)_3{\cdot}6H_2O$ as raw materials, in conjunction with the acid-catalyzed sol–gel method and supercritical drying, to prepare YSZ aerogels. They studied the influence of the sintering temperature on the crystalline form of the aerogels. Their YSZ aerogels changed from an amorphous phase to t-$ZrO_2$ at 455 °C, which remained stable at 1200 °C. After cooling to room temperature, the aerogel remained as t-$ZrO_2$.

The stabilization effects of $Y_2O_3$ on $ZrO_2$ aerogels are limited at temperatures less than 1200 °C; furthermore, crystal particles grow and a phase transition occurs when the temperature exceeds 1200 °C. For this reason, researchers have developed $ZrO_2$ aerogels that are stable at high temperatures.

Yu et al. (2020) used $ZrOCl_2$·8H2O, $Y(NO_3)_3{\cdot}6H_2O$, and rare-earth (RE) nitrates (RE = La, Ce, Gd, or Yb) as raw materials, in conjunction with sol-coagulation and supercritical drying, to prepare bulk RE-YSZ aerogels doped with RE elements. The phase stability of La-doped La–YSZ aerogels was improved compared with that of RE-YSZ aerogels doped with other RE elements. The RE-YSZ aerogels retained their stability without obvious microcracks between 30 °C and 1200 °C. The tetragonal phase of La-YSZ aerogels remained stable after heating at 1300 °C for 2 h.

### 1.3.2 Binary Composite Oxides

Preparing aerogels by adding a binary composite is currently the most commonly used approach to mitigate the decreased specific surface area of modified $ZrO_2$ aerogels caused by heating at temperatures higher than 200 °C.

Ren et al. (2015) used ZrO(NO3)2·5H2O as the zirconium source, and formamide (drying control) and tetraethoxysilane (surface modification), in conjunction with the sol-gel method and atmospheric drying, to produce $ZrO_2$-$SiO_2$ binary composite aerogels with a core-shell structure (Fig. 1.3). The specific surface area and density of the resulting aerogel were 619 $m^2$/g and 0.202 g/$cm^3$, respectively. After heating at 1000 °C, the amorphous state of the aerogel transformed into a tetragonal phase, whereas the crystallization was inhibited.

Zu et al. (Zou et al. 2016; Zu et al. 2014) used Zr(OC4H9)4 and Si(OC2H5)4 as raw materials, in conjunction with the sol–gel method and chemical liquid deposition (CLD), to produce $ZrO_2$/$SiO_2$ block composite aerogels. By depositing a layer of $SiO_2$ on the surface of $ZrO_2$ gel nanoparticles, they inhibited the crystal growth and phase transition of $ZrO_2$ at high temperatures. In so doing, they also prevented structural collapse and the corresponding particle size deterioration of the aerogel. The $ZrO_2$/ $SiO_2$ composite aerogel remained in the tetragonal phase after heating to 1000 °C for 2 h, with a specific surface area of 186 $m^2$/g. Figure 1.4 shows the specific surface areas of a CLD-modified $ZrO_2$ aerogel and a traditional $ZrO_2$-based aerogel after calcination at various temperatures. The former aerogel exhibited a larger specific surface area than that of the latter at 600 °C to 1000 °C.

Gao et al. (2018) used $ZrOCl_2$ as the precursor and Na2SiO3 as a gel initiator, in conjunction with co-hydrolysis, to prepare $ZrO_2$ aerogels coated with $SiO_2$ nanoshells. The introduced $SiO_2$ nanoshell layer, which served as a boundary enhancement addictive of the $ZrO_2$ particles, strengthened the connection of the $ZrO_2$ nanoparticles to form a massive aerogel and also considerably reduced the agglomeration of the $ZrO_2$ aerogel at high temperatures.

Li et al. (2014) used $ZrOCl_2$·8H2O and ethyl orthosilicate as raw materials and PO as a gel accelerator to prepare $SiO_2$-modified $ZrO_2$ aerogels. The $SiO_2$ coating layer inhibited diffusion, nucleation, and growth of $ZrO_2$—which greatly inhibited

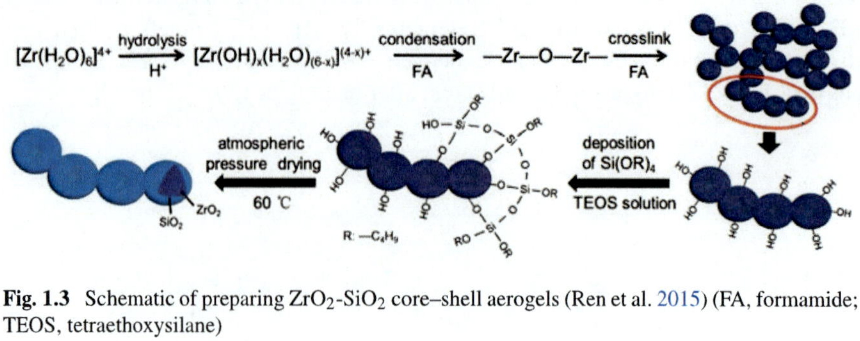


**Fig. 1.3** Schematic of preparing $ZrO_2$-$SiO_2$ core–shell aerogels (Ren et al. 2015) (FA, formamide; TEOS, tetraethoxysilane)

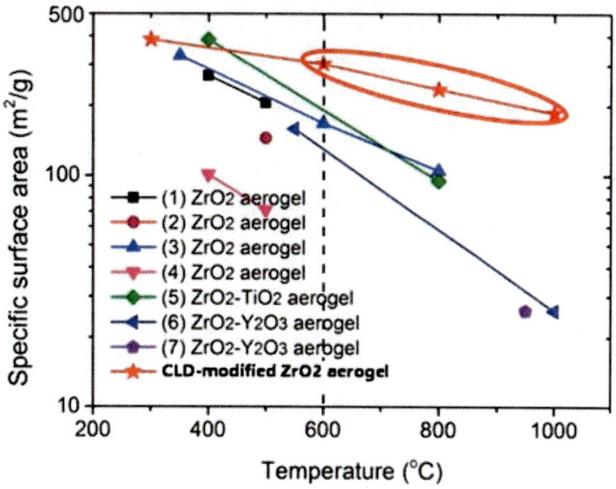


**Fig. 1.4** Comparison of the specific surface areas of a $ZrO_2$ aerogel modified by chemical liquid deposition and a traditional $ZrO_2$ aerogel at various temperatures (Zu et al. 2014)

growth of the crystal grains. Specifically, the size of the crystal particles was only 3–5 nm after heating at 1000 °C. They then used FeCl3 (Xiong et al. 2014) and hexamethyldisilazane (He et al. 2016) to modify the gel network structure; in so doing, they obtained a modified $ZrO_2$–$SiO_2$ composite aerogel that exhibited high-temperature stability as well as an improved specific surface area.

Doping $Al_2O_3$ or TiO2 in the binary composite aerogel is another method to improve the specific surface area and high-temperature stability of $ZrO_2$ aerogels. Jung et al. (2018) used zirconium n-propoxide and aluminum tri-sec-butoxide as raw materials, in conjunction with the sol–gel method and atmospheric drying, to prepare $ZrO_2$–$Al_2O_3$ composite aerogels with a Zr–O–Al structure. They investigated the influence of the aluminum tri-sec-butoxide content on the structural and high-temperature stability of $ZrO_2$–$Al_2O_3$ composite aerogels. When the molar ratio of aluminum tri-sec-butoxide to zirconium n-propoxide was 0.75, the specific surface area (91.9 $m^2$/g) of $ZrO_2$–$Al_2O_3$ aerogels at an annealing temperature of 800 °C was greater than that of a pure $ZrO_2$ aerogel (56.6 $m^2$/g).

Lian et al. (2015) used tetrabutyl zirconate and aluminum sec-butoxide as raw materials, in conjunction with ethanol supercritical drying, to produce $ZrO_2$–$Al_2O_3$ composite aerogels. Incorporating the acetone–aniline in-situ water-generation method in the preparation improved the formability of the resulting aerogels. The specific surface area of the resulting aerogel was 558 $m^2$/g, and remained at 129 $m^2$/g after heating to1000°C.

Wan et al. (2004) used $Ti(SO_4)_2$ and $ZrOCl_2$ as raw materials to prepare TiO2–$ZrO_2$ composite aerogels. The specific surface area of the aerogels remained at 95 $m^2$/g when the calcination temperature was 800 °C.

## 1.4 Appliances of $ZrO_2$ Aerogels

$ZrO_2$ aerogels exhibit excellent thermal and chemical stability catalytic performance, ultra-low density, low thermal conductivity, high porosity, and high specific surface area. Accordingly, researchers widely use such aerogels in thermal insulation, and catalysis, adsorption, and batteries.

### *1.4.1 Thermal Insulation Materials*

There are three methods of heat transfer in aerogel materials: (1) solid-phase, (2) gas-phase, and (3) radiation (Wu et al. 2015). Because the pore size of $ZrO_2$ aerogels (2–50 nm) is smaller than the mean free path of the main components of air (i.e., O2 and N2 molecules; ca. 70 nm), the collision probability between the gas molecules in the aerogel is low and the effect of convective heat transfer in the aerogel is negligible. $ZrO_2$ aerogels have a high porosity and low density, which greatly reduces the proportion of solid-phase heat conduction, imparting $ZrO_2$ aerogels with low thermal conductivity. To date, most research on $ZrO_2$ aerogels focuses on preparation methods and the corresponding stability of the resulting aerogel, whereas there are comparatively few studies on corresponding thermal insulating applications, especially at high temperatures (Liu et al. 2019b; Yoon et al. 2019).

Yoon et al. (Yoon et al. 2019) prepared YSZ aerogels by the sol–gel method and supercritical $CO_2$ drying. Then they mixed the resulting aerogel with an inorganic binder, and coated the product on the surface of a gas turbine blade (Fig. 1.5). The researchers studied the thermal insulation performance of the YSZ aerogel by measuring the thermal conductivity of the YSZ aerogel and the surface temperature distribution of the thermal barrier coating. The thermal conductivity of the YSZ aerogel at 1000 °C was 0.212 W $m^{-1}$ $K^{-1}$, which was much lower than that of YSZ materials prepared by other methods (0.5–2.36 W $m^{-1}$ $K^{-1}$). When the temperature was between 300 and 700 °C, the surface temperature of the YSZ aerogel thermal barrier coating was 30% to 50% lower than that of a YSZ thermal barrier coating prepared by air plasma spraying.

### *1.4.2 Catalysts and Catalyst Supports*

$ZrO_2$ aerogels are common catalysts and catalyst supports. $ZrO_2$ aerogels exhibit acidity, alkalinity, and redox activity concomitantly (Tanabe 1985); furthermore, $ZrO_2$ aerogels have a small pore size and high specific surface area, which increases the dispersion rate of active components and correspondingly improves the catalytic activity, thermal stability, and anti-poisoning performance (Lee et al. 2015). One can group reported studies on $ZrO_2$ aerogel catalysts as follows:

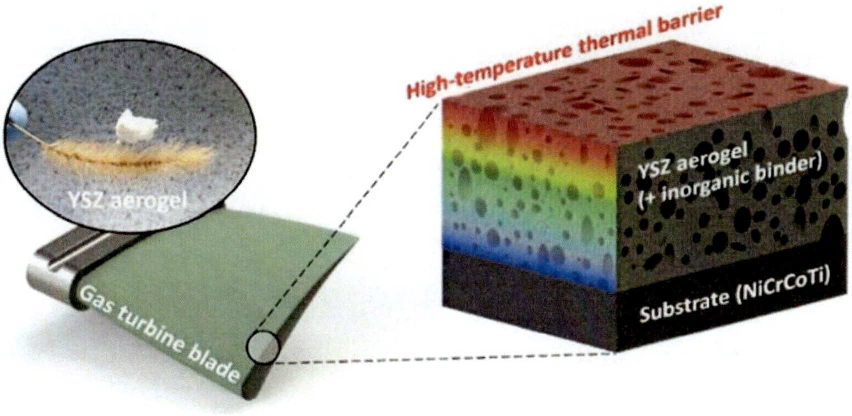


**Fig. 1.5** Thermal barrier coating of a $Y_2O_3$-stabilized $ZrO_2$ (YSZ) aerogel used to coat a gas turbine blade (Yoon et al. 2019)

(1) Expanding the applications of $ZrO_2$ aerogels as catalyst supports by doping with active metals (such as copper, iron, and platinum). For example, a Fe/$ZrO_2$ composite aerogel can impart high catalytic activity for complete oxidation of toluene in water (Ismail et al. 2020). One can use a Pt/$ZrO_2$ aerogel catalyst for CO hydrogenation to synthesize formate (Kalies et al. 2000), and a Cu/$ZrO_2$ composite aerogel to catalyze hydroxylation of phenol (Li 2019) and hydrogenation of CO and $CO_2$ (Sun and Sermon 1993; Sermon et al. 1997).

(2) Improve the surface acidity and catalytic activity of $ZrO_2$ aerogels through modification of the surface of the aerogel—by using anions such as sulfate, phosphate, and tungstate (Yadav and Nair 1999; Chakhari et al. 2015; Boyse and Ko 1996). For example, researchers used a SO42 − /$ZrO_2$ aerogel catalyst as a solid super-acid that exhibited good catalytic activity toward synthesis of methyl stearate (Saravanan et al. 2016).

### *1.4.3 Adsorbents*

$ZrO_2$ aerogels are acid–base amphoteric nanoporous materials that have a large specific surface area and high porosity. Because of their high adsorption efficiency and adsorption selectivity, one can use such aerogels to adsorb organic pollutants, phosphorylated peptides, and heavy-metal ions.

Zhao et al. (2016) prepared $SiO_2$–$ZrO_2$ composite aerogels by phase separation. The resulting aerogel exhibited good adsorption performance for simulated Ce(IV) ions in a radioactive waste liquid. Because the atomic radius of Zr is close to that of the radionuclide, the adsorbed powder forms a stable solidified body after high-temperature calcination, which provides a safe and effective means of treating waste streams that contain the radionuclide.

Zhang et al. (2011) prepared $ZrO_2$ aerogels by using a hydrophilic mesoporous material. The resulting aerogel had good selectivity and a high enrichment capacity for phosphorylated peptides.

Huang (2018) prepared $SiO_2$–$ZrO_2$ composite aerogels by the sol–gel method and atmospheric drying. They tested the adsorption of the resulting aerogel for two dyes: rhodamine B and methylene blue. The removal rate for the two dyes reached ca. 85% within the first 10 min; and the maximum adsorption capacity reached 177.7 and 63.13 mg/g, respectively.

## 1.5 Conclusions and Prospects

Advances in $ZrO_2$ aerogel preparation and drying, modification, and applications were reviewed in the study. The widely used preparation methods include (1) zirconium alkoxide hydrolysis, (2) electrolysis, (3) precipitation, (4) alcohol–water solution heating, (5) PO addition, (6) inorganic dispersion sol–gel, (7) direct sol–gel, and (8) organic acid assisted sol–gel. Among these methods, the electrolysis, PO addition, and inorganic dispersion sol–gel methods have more advantages in terms of environmental friendliness and ease of scale-up. Although research on applications of the direct sol–gel and organic acid-assisted sol–gel methods remain under-developed, these two methods have great potential in future preparations of aerogels because of their low toxicity and low safety risk.

The widely used drying methods include (1) supercritical drying, (2) freeze-drying, and (3) atmospheric drying. Atmospheric drying is the most active field of research because of its low cost and compatibility with conventional atmospheric pressures. The widely used modification methods include (1) doping modification and (2) adding a binary composite. The latter is comparatively more common because $ZrO_2$ aerogels prepared via this method exhibit improved stability at high temperatures compared with corresponding aerogels prepared by doping. Because of their acid–base duality, redox properties, high specific surface areas, high porosities, and high level of thermal and chemical stability, researchers have applied $ZrO_2$ aerogels as follows: (1) thermal insulation, (2) catalysts and catalyst support, and (3) adsorbents.

Room for improvement is as follows: (1) Supercritical drying remains the most common preparation process for $ZrO_2$ aerogels, yet this process requires high temperatures and pressures. Such conditions are dangerous and expensive, and impart difficulties to producing aerogels on a large scale. Therefore, researchers should develop atmospheric drying technology that is compatible with mild conditions by utilizing simple equipment and rapid drying. (2) Applications of pure $ZrO_2$ aerogels remain limited because of such aerogels' poor mechanical performance and thermal conductivity at high temperatures. Therefore, some modifications to the composite and structure of $ZrO_2$ are required to improve the mechanical performance of $ZrO_2$. (3) Most of the microstructure units in the aerogels are $ZrO_2$ nanoparticles, which have poor thermal stability and readily sinter when the temperature is higher than

1000 °C. Accordingly, microstructures such as one-dimensional nanowires and two-dimensional nanosheets should be active areas of future research. (4) The replacement of toxic and harmful organic zirconium alkoxides and PO gel accelerators with cheap, low-toxic inorganic zirconium salt precursors and gel accelerators will enhance the environmental sustainability of preparing $ZrO_2$ aerogels.

**Acknowledgements** This work was supported in part by the General Program of Science and Technology Innovation Committee of Shenzhen (Grant No. JSGG20191129110221063), and in part by the Key Technology Research and Application Program of Science and Technology Innovation Committee of Shenzhen (Grant No. GJHZ20200731095611035).

## References

Bai L, Ma H, Gao C, et al. (2006) $ZrO_2$ aerogel prepared by heating of alcohol-aqueous salt solution. J Mol Catalysis (China) 20(6):539–544

Bangi UKH, Park HH (2018) Evolution of textural characteristics of surfactant-mediated mesoporous zirconia aerogel powders prepared via ambient pressure drying route. Int Nano Lett 8(3):221–228

Bangi UKH, Lee K-Y, Maldar NMN, et al. (2019) Synthesis and properties of metal oxide aerogels via ambient pressure drying. J Nanosci Nanotechnol 19(3):1217–1227

Bangi UKH, Patil B, Pawar RC, et al. (2020) Influence of various sol–gel parameters on the physico-chemical properties of sulfuric acid chelated zirconia aerogels dried at ambient pressure. Macromol Symp 393(1):2000025 (2020)

Bedilo AF, Klabunde KJ (1997) Synthesis of high surface area zirconia aerogels using high temperature supercritical drying. Nanostruct Mater 8(2):119–135

Bianchi, D; Gass, J-L; Khalfallah, M; et al. Intermediate species on zirconia supported methanol aerogel catalysts: I. State of the catalyst surface before and after the adsorption of hydrogen. Appl Catal A: General 101(2):297–315 (1993)

Boyse RA, Ko EI (1996) Preparation and characterization of zirconia-phosphate aerogels. Catal Lett 38(3):225–230

Cao Y, Hu J-C, Hong Z-S et al (2002) Characterization of high-surface-area zirconia aerogel synthesized from combined alcohothermal and supercritical fluid drying techniques. Catal Lett 81(1):107–112

Chakhari S, Younes MK, Rives A et al (2015) Effect of the doping agent nature on the characteristic and catalytic properties of aerogel zirconia catalysts doped with sulfate groups or heteropolytungstic acid. Mater Res Bull 72:35–42

Chao X, Yuan W, Shi Q et al (2016) Improvement of thermal stability of zirconia aerogel by addition of yttrium. J Sol-Gel Sci Technol 80(3):667–674

Chervin CN, Clapsaddle BJ, Chiu HW et al (2006) Role of cyclic ether and solvent in a non-alkoxide sol−gel synthesis of yttria-stabilized zirconia nanoparticles. Chem Mater 18(20):4865–4874

Du A, Zhou B, Shen J et al (2009) Monolithic copper oxide aerogel via dispersed inorganic sol–gel method. J Non-Cryst Solids 355(3):175–181

Du A, Zhou B, Zhang Z et al (2013) A special material or a new state of matter: a review and reconsideration of the aerogel. Materials 6(3):941–968

Gao H, Zhang Z, Shi Z et al (2018) Synthesis of high-temperature resistant monolithic zirconia-based aerogel via facile water glass assisted sol–gel method. J Sol-Gel Sci Technol 85(3):567–573

Garvie RC (1978) Stabilization of the tetragonal structure in zirconia microcrystals. J Phys Chem 82(2):218–224

Gash AE, Tillotson TM, Satcher JH et al (2001a) Use of epoxides in the sol−gel synthesis of porous iron(III) oxide monoliths from Fe(III) salts. Chem Mater 13(3):999–1007
Gash, A.E; Tillotson, T.M; Satcher, Jr.J.H; et al. New sol–gel synthetic route to transition and main-group metal oxide aerogels using inorganic salt precursors. Journal of Non-Crystalline Solids, 2001b, 285(1): 22–28.
Guan Y, Jiang Y, Feng J et al (2017) Research progress on inorganic fiber reinforced silica aerogel insulation composites. Materials Reports 31(S1):429–434
Guo X, Yan L, Yang H et al (2011a) Synthesis of zirconia aerogels by ambient pressure drying with propylene oxide addition. Acta Phys Chim Sin 27(10):2478–2484
Guo X, Yang H, Yan L, et al. (2011b) Preparation method of ZrO2 aerogel, CN101961626A [P/OL]. 2011b–02–02
He J, Li X, Su D et al (2016) Super-hydrophobic hexamethyl-disilazane modified ZrO2–$SiO_2$ aerogels with excellent thermal stability. J Mater Chem A 4(15):5632–5638
Hu J, Cao Y, Deng J (2001) A simple alcohothermal synthetic route to high surface area zirconia aerogel. Chem Lett 30(5):398–399
Hu L, He R, Lei H et al (2019) Carbon aerogel for insulation applications: a review. Int J Thermophys 40(4):39
Huang Y-Y, Zhao B-Y, Xie Y-C (1998) Preparation of zirconia-based acid catalysts from zirconia aerogel of tetragonal phase. Appl Catal A 172(2):327–331
Huang G (2018) The atmospheric drying preparation and adsorption properties of $SiO_2$-$ZrO_2$ composite aerogels, Doctoral Thesis, Harbin Institute of Technology, Harbin (China), 2018
Hurwitz FI, Rogers RB, Guo H et al (2020) Phase development and pore stability of yttria- and ytterbia-stabilized zirconia aerogels. J Am Ceram Soc 103(12):6700–6711
Ismail R, Arfaoui J, Ksibi Z et al (2020) Effect of the iron amount on the physicochemical properties of Fe– ZrO2 aerogel catalysts for the total oxidation of Toluene in the presence of water vapor. J Porous Mater 27(6):1847–1852
Jung H-N-R, Han W, Cho HH, et al. (2017) Effect of cationic and non-ionic surfactants on the microstructure of ambient pressure dried zirconia aerogel. Mater Express 2017, 7(4):291–298
Jung, H-N-R, Parale VG, Kim T, et al. (2018) Zirconia-based alumina compound aerogels with enhanced mesopore structure. Ceramics Int 44(9):10579–10584
Kalies H, Pinto N, Pajonk GM et al (2000) Hydrogenation of formate species formed by CO chemisorption on a zirconia aerogel in the presence of platinum. Appl Catal A 202(2):197–205
Khajavi P, Babaluo AA, Tavakoli A et al (2014) Stabilization of the metastable tetragonal phase in zirconia nanopowders synthesized via polyacrylamide gel method. Ind Eng Chem Res 53(1):164–172
Kistler SS (1931) Coherent expanded aerogels and jellies. Nature 127(3211):741–741
Le X, Zhao C, Chen J et al (2018) Preparation of cellulose/$ZrO_2$ aerogels by non-supercritical drying. Metal World 06:20–24
Lecomte A, Blanchard F, Dauger A et al (1998) Synthesis and sintering of zirconium oxide aerogel. J Non-Cryst Solids 225:120–124
Lee Y, Choi J-W, Suh DJ et al (2015) Ketonization of hexanoic acid to diesel-blendable 6-undecanone on the stable zirconia aerogel catalyst. Appl Catal A 506:288–293
Li X, Jiao Y, Ji H et al (2013) The effect of propylene oxide on microstructure of zirconia monolithic aerogel. Integr Ferroelectr 146(1):122–126
Li X, Wang Q, Ji H et al (2014) High temperature stability of $ZrO_2$ aerogel modified by $SiO_2$. Aerospace Mater Technol 44(01):65–68
Li Z (2019) Preparation of zirconia-Fe and Cu aerogel catalysts and catalytic activity for phenol hydroxylation, Doctoral Thesis, Changchun University of Technology, Changchun (China), 2019
Li, W; Song, Y; Huang, G; et al. Method for preparing zirconia zerogel by secondary gel method combined with atmospheric drying, CN105271403A [P/OL]. 2016–01–27
Lian Y, Shen J, Zu G et al (2015) Preparation and Characterization of ZrO_2–Al_2O_3 Composite Aerogels. J Chin Ceramic Soc 43(11):1656–1662

Liang L, Hou X, Wu D (2005) Synthesis of ZrO_2 aerogels by supercritical CO_2 fluid drying technology. Mater Sci Technol 13(5):106–109
Liu B, Liu X, Zhao X et al (2019a) High-strength, thermal-stable $ZrO_2$ aerogel from polyacetylacetonatozirconium. Chem Phys Lett 715:109–114
Liu B, Gao M, Liu X et al (2019b) Thermally Stable Nanoporous ZrO2/$SiO_2$ Hybrid Aerogels for Thermal Insulation. ACS Applied Nano Materials 2(11):7299–7310
Mu, Y. Synthesis of Zirconia Aerogels by Ambient Pressure Drying, Doctoral Thesis, Shanghai Institute of Technology, Shanghai (China), 2017.
Nadargi DY, Latthe SS, Venkateswara RA (2009) Effect of post-treatment (gel aging) on the properties of methyltrimethoxysilane based silica aerogels prepared by two-step sol–gel process. J Sol-Gel Sci Technol 49(1):53–59
Pierre AC, Pajonk GM (2002) Chemistry of aerogels and their applications. Chem Rev 102(11):4243–4266
Ren J, Cai X, Yang H et al (2015) Preparation and characterization of high surface area ZrO2 aerogel modified by $SiO_2$. J Porous Mater 22(4):973–978
Saravanan K, Tyagi B, Bajaj HC (2016) Nano-crystalline, mesoporous aerogel sulfated zirconia as an efficient catalyst for esterification of stearic acid with methanol. Appl Catal B 192:161–170
Schäfer H, Milow B, Ratke L (2013) Synthesis of inorganic aerogels via rapid gelation using chloride precursors. RSC Adv 3(35):15263–15272
Sermon PA, Self VA, Sun Y (1997) Doped-ZrO2 aerogels: Catalysts of controlled structure and properties. J Sol-Gel Sci Technol 8(1):851–856
Shen L (2017) Research on thermal stability and molding process of ZrO_2 aerogel. Inorganic Chemicals Industry 49(05):30–33
Srikanth, C; Madhu, G. (2017) Synthesis, characterization and properties evaluation of $ZrO_2$ and Its Composites–A Review, Synthesis, Characterization and Properties Evaluation of $ZrO_2$ and Its Composites–A Review. Int Conf Adv Thermal Syst, Mater Design Eng (ATSMDE2017)
Suh DJ, Park T-J (1996) Sol−Gel Strategies for Pore Size Control of High-Surface-Area Transition-Metal Oxide Aerogels. Chem Mater 8(2):509–513
Sui R, Charpentier P (2012) Synthesis of metal oxide nanostructures by direct sol-gel chemistry in supercritical fluids. Chem Rev 112(6):3057–3082
Sui R, Rizkalla AS, Charpentier PA (2006) Direct Synthesis of Zirconia Aerogel Nanoarchitecture in Supercritical $CO_2$. Langmuir 22(9):4390–4396
Sun Q, Zhang Y, Deng J et al (1997) A novel preparation process for thermally stable ultrafine tetragonal zirconia aerogel. Appl Catal A 152(2):L165–L171
Sun Y, Sermon PA (1993) Carbon monoxide hydrogenation over $ZrO_2$ and Cu/$ZrO_2$. J Chem Soc, Chem Commun (16):1242–1244
Tanabe K (1985) Surface and catalytic properties of ZrO2. Mater Chem Phys 13(3):347–364
Tao Y, Endo M, Kaneko K (2008) A review of synthesis and nanopore structures of organic polymer aerogels and carbon aerogels. Recent Pat Chem Eng 1(3):192–200
Teichner SJ, Nicolaon GA, Vicarini MA et al (1976) Inorganic oxide aerogels. Adv Coll Interface Sci 5(3):245–273
Torres-Rodríguez J, Kalmár J, Menelaou M et al (2019) Heat treatment induced phase transformations in zirconia and yttria-stabilized zirconia monolithic aerogels. The Journal of Supercritical Fluids 149:54–63
Veselovskaya JV, Derevschikov VS, Shalygin AS et al (2021) $K_2CO_3$-containing composite sorbents based on a $ZnO_2$ aerogel for reversible $CO_2$ capture from ambient air. Micropor Mesopor Mater 310:110624
Wan Y, Ma J, Zhou W et al (2004) Preparation of titania–zirconia composite aerogel material by sol–gel combined with supercritical fluid drying. Appl Catal A 277(1):55–59
Wang Y (2014) Research progress of high-performance aerogel thermal insulation materials. Aerodyn Missile J 03:90–94
Wang X, Li C, Shi Z et al (2018) The investigation of an organic acid assisted sol–gel method for preparing monolithic zirconia aerogels. RSC Adv 8(15):8011–8020

Wu Z, Zhao Y, Liu D (2004) Preparation and characterization of ZrO_2 aerogel. J Funct Mater 35(3):389–391

Wu X, Cui S, Wang L et al (2015) Advance in research of high temperature resistant aerogel used as insulation material. Mater Rev 29(09):102–108

Xiao S, Zhou B, Wan H et al (2008) Monolithic lithium-based aerogels via dispersed inorganic sol-gel method. Atomic Energy Sci Technol 42(S1):21–25

Xiong R, Li X, Ji H et al (2014) Thermal stability of ZrO2–$SiO_2$ aerogel modified by Fe(III) ion. J Sol-Gel Sci Technol 72(3):496–501

Yadav GD, Nair JJ (1999) Sulfated zirconia and its modified versions as promising catalysts for industrial processes. Microporous Mesoporous Mater 33(1):1–48

Yoon S, Han GD, Jang DY et al (2019) Fabrication of yttria-stabilized zirconia aerogel for high-performance thermal barrier coating. J Alloy Compd 806:1430–1434

Yu Q, Zhou B, Zhang Z et al (2014) Antimony-doped tin oxide aerogel based on epoxide additional method. Acta Phys Chim Sin 30(03):500–507

Yu C, Yu Y, Zhao M et al (2020) Phase stability and grain growth of rare earth oxide $RE_2O_3$(RE=La, Ce, Gd, Yb)doped yttrium stabilized zirconia aerogel. Bull Chin Ceramic Soc 39(05):1620–1626

Yu K, Suo H, Cui S, et al. (2018) Advances in research of high temperature resistant adiabatic carbide aerogel materials. Mod Chem Ind 38(03):47–51+53

Zhang L, Xu J, Sun L et al (2011) Zirconium oxide aerogel for effective enrichment of phosphopeptides with high binding capacity. Anal Bioanal Chem 399(10):3399–3405

Zhang Z, Gao Q, Liu Y et al (2015) A facile citric acid assisted sol–gel method for preparing monolithic yttria-stabilized zirconia aerogel. RSC Adv 5(102):84280–84283

Zhao Z, Chen D, Jiao X (2007) Zirconia Aerogels with high surface area derived from sols prepared by electrolyzing zirconium oxychloride solution: comparison of aerogels prepared by freeze-drying and supercritical $CO_2$(l) extraction. J Phys Chem C 111(50):18738–18743

Zhao J, Ding G, Meng C et al (2016) Preparation of SiO_2-ZrO_2 aerogels and the adsorption performance for cerium. Rare Metal Mater Eng 45(S1):266–270

Zhong L, Chen X, Song H et al (2014) Synthesis and characterization of zirconia aerogels by hydrothermal-gel method. Aerospcae Mater Technol 44(1):72–74

Zhu J, Liu F, Zhou C et al (2015) Research progress of zirconia aerogel. Adv Ceramics 36(03):30–36

Zou W, Shen J, Zu G et al (2016) Preparation and characterization of heat-resistant zirconia/silica composite aerogel. J Nanjing Univ Technol (Natural Science Edition) 38(02):42–46

Zu G, Shen J, Wang W et al (2014) Robust, highly thermally stable, core-shell nanostructured metal oxide aerogels as high-temperature thermal superinsulators, adsorbents, and catalysts. Chem Mater 26(19):5761–5772

# Chapter 2
# Survey and Research on Mass Participation in Sponge City Construction in Jinhua City

**Linqi Gu, Jiang Cheng, and Yong Liang**

**Abstract** In June this year, Jinhua City was officially selected as the second batch of 25 demonstration cities for systematic and full-field promotion of sponge city construction in the country, and strive to construct the first-class sponge city construction demonstration city in China within three years. In order to better construct Jinhua sponge city demonstration city, this paper surveys and researches the participation of the masses in Jinhua sponge city through online questionnaires. The survey found that there are problems such as knowing little about sponge city, as well as insufficient understanding and participation among the masses. Therefore, based on the survey and research of sponge city in Jinhua, this paper provides suggestions for the construction of the sponge city construction demonstration city in Jinhua, with a view to providing some reference for the construction of sponge city in Jinhua and other regions.

**Keywords** Stormwater management · Sponge city construction demonstration city · Publicity · Awareness · Participation

**CCS Concepts** Applied computing · Physical sciences and engineering · Earth and atmospheric sciences · Environmental sciences

---

L. Gu · J. Cheng (✉) · Y. Liang
Zhejiang Province Institute of Architectural Design and Research, Hangzhou, China
e-mail: sklec@qq.com

L. Gu
e-mail: 294969245@qq.com

Y. Liang
e-mail: 148100303@qq.com



P. Mendonça et al. (eds.), *Proceedings of 2023 International Conference on Green Building*, Environmental Science and Engineering,
https://doi.org/10.1007/978-3-031-43478-5_2

## 2.1 Introduction

Sponge city, is a new generation of urban stormwater management concept (Geiger et al. 2015), refers to the city that can be like a sponge, adapting to environmental changes and responding to natural disasters brought about by rainwater has good resilience, can also be called "water resilient city". The international term for this is "Construction of Rainwater System with Low Impact Development", which absorbs, stores, infiltrates and purifies water when it rains, and releases and uses the stored water when needed to achieve free movement of rainwater in the city. However, with the rapid development of cities, the problems of stormwater runoff pollution, water scarcity, flooding and loss of aquatic habitats have become increasingly prominent (Wang et al. 2015; Che et al. 2015a), especially in the case of the "July 20 Heavy rainstorm in Zhengzhou", which once became the focus of the national media and once again attracted close attention to the issue of urban flooding.

In April this year, the Ministry of Housing and Urban–Rural Development released "Sponge 20 strips", which demands for the solid promotion of sponge city construction and the enhancement of urban flood control and drainage capacity. There is no doubt that sponge city, as a new manifestation of innovation and development in rainwater management, achieve a leapfrog shift from "engineering water management" to "ecological water management", alleviating urban flooding at source, reducing rainwater runoff pollution, saving water resources and restoring the urban water ecology (Liao et al. 2016; Che et al. 2015b).

In June this year, Jinhua City was among the second batch of 25 demonstration cities in the "14th Five-Year Plan" to systematically promote the construction of sponge city in the country, and received 900 million central financial support. Sponge city is the highest priority in the construction of the national ecological civilization demonstration city, but research has found that the mass lack of understanding of sponge city construction in Jiaxing has hindered the progress of the project to a certain extent (Zhu et al. 2017). To focus on constructing the first-class sponge city construction demonstration city in China, this paper surveys and researches the participation of the masses in the construction of sponge city in Jinhua City, which is not only of great significance for Jinhua City to construct a national demonstration city better and faster, but also hopes to accumulate experience for the construction of sponge city in Zhejiang Province and even the whole country, to guide the construction of sponge city that really satisfies the masses.

## 2.2 Methods and Results

### 2.2.1 Duration and Scope of the Survey

The survey lasted for 14 days from June 1 to 14, 2022. The scope of the survey was the masses in Jinhua City, and the survey adopted the online questionnaire method. All questionnaires were filled out on a voluntary basis, once per person, anonymously, for statistical analysis only and for no other purpose, without revealing personal information, and with no right or wrong answers.

### 2.2.2 Survey Contents

The survey consisted of 11 questions, the details of which are shown in Table 2.1.

**Table 2.1** The contents of the mass questionnaire for the construction of sponge city in Jinhua City

| Order | Survey contents |
|---|---|
| 1 | Do you live in Jinhua City? A Yes; B No |
| 2 | Your gender? A Male; B Female |
| 3 | Your age? A Under 20 years of age; B 20–40 years old; C 41–60 years old; D Over 60 years old |
| 4 | Your occupation? A Civil servant, enterprise or institution employee; B Company employee, freelancer; C Student; D Retired person; E Other |
| 5 | Your education? A Senior secondary school and below; B Junior college and undergraduate college; C Postgraduate |
| 6 | Do you know about sponge city? A Know; B Unknow |
| 7 | Do you agree that sponge city has the functions of "light rain does not accumulate water, heavy rain does not waterlog, water body is not black and smelly, and heat island is alleviated"? A Yes; B No; C Not sure |
| 8 | Yanweizhou and Huhaitang Parks are excellent sponge city projects in Zhejiang Province, do you know? A Yes; B No |
| 9 | Your evaluation of the effectiveness of sponge city construction in Yanweizhou and Huhaitang Parks are: A Excellent; B Good; C General |
| 10 | Have you been involved in projects or activities related to sponge city? A Yes; B No |
| 11 | Do you support the construction of the sponge city construction demonstration city in Jinhua? A Yes; B No |

*Note* For the first question, if you choose "A Yes", continue to answer questions 02–11; if you choose "B No", you will skip directly to the "Thank you" page, indicating that the subject is not within the scope of the survey, which is invalid

### 2.2.3 Survey Results

A total of 660 copies (2574 views) of the questionnaire were collected, and 660 people participated in this survey, of whom 34 were not in Jinhua City and 626 were in Jinhua City. Therefore, the survey results of questions 2–10 were based on the votes of 626 people in Jinhua City.

Among the subjects (626 people within Jinhua City) of this survey, 54.63% of subjects were male, 45.37% were female, the age group was mainly concentrated in the 20–60 years old group, the occupation was mainly concentrated in civil servants, employees of enterprises and institutions and company employees, freelancers, the education level was mainly concentrated in junior college and undergraduate college, 67.41% knew about sponge city, 66.45% agreed with the concept of sponge city, 65.97% knew about the construction of sponge city in Yanweizhou and Huhaitang Parks, 61.98% evaluated the construction of sponge city in Yanweizhou and Huhaitang Parks as excellent, 18.37% have participated in sponge city related projects or activities, and 96.49% supported the construction of systematic and full-field promotion of the sponge city construction demonstration city in Jinhua (Fig. 2.1).

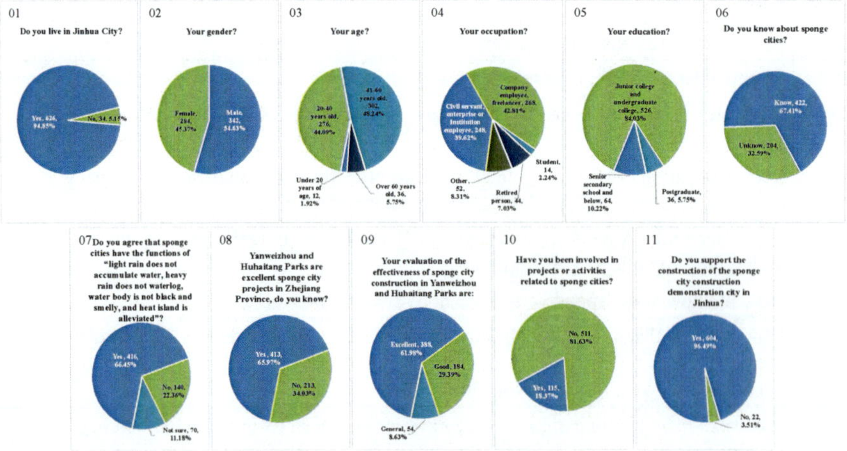


**Fig. 2.1** Figure of questionnaire results. *Note* The figure is marked in the form of "X, X1, X2", where "X" represents the content of the option, "X1" indicates the number of people who choose the item, and "X2" indicates the percentage of the number of people who chose the item to the number of subjects (626 people) in Jinhua City

## 2.3 Analysis of Mass Participation in Sponge City Construction

Based on the analysis of the results of this survey, it shows that the following problems mainly exist in the process of mass participation in sponge city in Jinhua.

### *2.3.1 Knowing Little About Sponge City Among the Masses*

After survey and analysis, although Jinhua City has achieved certain results in the construction of sponge city, the publicity of sponge city construction was still insufficient (She et al. 2021). For instance: the masses do not know much about sponge city, the concept of constructing sponge city is still vague, mainly because only 67.41% of subjects know sponge city, there are some people who do not agree with the concept of sponge city, 34.03% of subjects do not know Yanweizhou and Huhaitang Parks are excellent sponge city projects in Zhejiang Province. These survey results show that some people do not understand "what is a sponge city" and have a negative attitude towards the concept of sponge city. Through the above analysis, it is generally shown that the public is not very familiar with sponge city, and the government needs to increase the publicity of sponge city to let the masses know what is sponge city, so that the concept of "sponge city" can be deeply rooted in the hearts of the people, and really perceive the visible benefits brought by the sponge city.

### *2.3.2 Lack of Understand of Sponge City Among the Masses*

Through surveys and analysis, we learned that nearly 70% people do not know and are not clear about "what is a sponge city", although government departments have introduced sponge city on different platforms such as the government official websites, WeChat official accounts, etc., the effect is not very good, the public's awareness rate is not high, and many people do not know about these platforms, or even have not read the relevant reports. This has resulted in the government working hard to promote sponge city, but the public may still not have effective access to sponge city related publicity knowledge, let alone knowing about these excellent provincial sponge city projects such as Yanweizhou and Huhaitang Parks, and will not be tangibly involved in sponge city construction. In the follow-up contact with the masses, it was found that some people think that sponge city is just "face projects", with large investments and long construction periods, bringing inconvenience to people's daily life, while the benefits are low and do not work at all when it rains heavily, as usual, "the ground is full of water". Through the above analysis, it is generally shown that the masses have one-sided understanding and lack of understanding of sponge city.

### 2.3.3 *Lack of Participation in Sponge City Among the Masses*

In the process of field survey, many people said that they were busy with work and could not or did not want to participate in it, and at the same time, they did not know how to participate or whether they had ever participated. From the survey results of question 10 of this online questionnaire, it can also be seen that the public's participation in the construction of sponge city is not high, and only 18.37% of them have participated in sponge city related projects or activities. At the same time, from the participation of this questionnaire, it can also be seen that the masses of low participation, the questionnaire collected a total of 660 copies, a total of 2574 views, the recovery rate of 25.64%, more than 70% of the people just browsing, did not effectively fill in, indicating that the public may not understand or attach importance to the sponge city, did not actively participate, and a very small number of people do not support Jinhua City to construct a systematic and full-field promotion of the sponge city construction demonstration city. Through the above analysis, it is generally indicated that the participation of the masses is not enough, they do not actively participate or have no interest in sponge city.

## 2.4 Suggestions for Increasing Mass Participation in the Construction of the Sponge City

### 2.4.1 *Strengthen Publicity*

In order to better publicize the sponge city, it is necessary to strengthen the publicity (Zhou et al. 2021). On the one hand, government departments, enterprises and institutions, schools, and street communities should let mass actively pay attention to the online channels of sponge city through offline publicity, such as the government's official websites and WeChat public accounts, and guide them to actively read the contents related to sponge city. On the other hand, in combination with offline experience, actively explore the publicity forms popular with the masses, for example, exhibit the sponge city model or publicity materials in the science museum, and introduce them in a language that the masses can understand, so as to win the understanding, support and cooperation of the public on the construction and transformation of the sponge city, so that the concept of the sponge city can be recognized by the masses.

### 2.4.2 *Reverse Negative Perceptions*

In order to better promote sponge city, negative perceptions must be reversed (Ding et al. 2019). Some people have a negative view of sponge city, believing that the construction of sponge city has no substantial use, but increases the investment and construction period, which has a bad impact on the daily life of the masses. In order to reverse negative perceptions, it is necessary for the public to understand the role of sponge city and what substantial benefits they can bring us. For example, through league construction or community activities, let the masses walk on the permeable pavement of the provincial excellent sponge city projects, such as Yanweizhou and Huhaitang Parks, after a rainstorm, to see if there is water on the pavement, walk around to see if their shoes are wet, and let the masses feel first-hand the visible benefits of the sponge city to their daily lives, in order to really reverse this one-sided understanding. With the support and understanding of the masses, the construction of sponge city can develop faster.

### *2.4.3 Solidly Promote Projects*

In order to construct sponge city better, we must solidly promote sponge city construction projects. The national strategy of sponge city construction is not only the responsibility of the government, but also requires the active participation of social forces, and every public should also be a real participant. On the one hand, government personnel themselves should strengthen the construction management and technological improvement of sponge city, and truly implement the construction of sponge city, instead of treating sponge city projects as "image projects" and "face projects" on paper. On the other hand, drawing on domestic and international experience, combined with the situation in Jinhua, give certain policy tendencies to the masses who participate in the construction of sponge city, while fully listening to public opinion and meeting the reasonable needs of the masses. For example, the Swiss government uses policies such as tax breaks and subsidized allowances to encourage people to construct water-saving houses and save water through rainwater recycling (Li 2021; Köster 2021). This not only solved the government in the sponge city construction alone, helpless situation, but also mobilized the power of the whole people to carry out the construction of the sponge city. In this way, sponge city construction can have a real practical effect and become a real project to benefit the people, enhance the sense of happiness, access and security.

## References

Che W et al (2015a) Considerations and discussions about Sponge city. South Archit 4:104–108 (in Chinese)

Che W et al (2015b) Explanation of Sponge city development technical guide: basic concepts and comprehensive goals. China Water Wastewater 31(8):1–5 (in Chinese)

Ding L et al (2019) Implementation of the Sponge city development plan in China an evaluation of public willingness to pay for the life cycle maintenance of its facilities. Cities 93:13–30

Geiger WF et al (2015) Sponge city and LID technology—vision and tradition. Landsc Archit Front 3(2):10–20

Köster S (2021) How the Sponge city becomes a supplementary water supply infrastructure. Water-Energy Nexus 4(4):35–37

Li H (2021) Characteristics of regional water resources circulation based on sustainable development and landscape design of Sponge city. J Environ Protect Ecol 22(2):627–634

Liao C et al (2016) Enlightenment of rainwater management in foreign countries to sponge city construction in China. Water Resourc Protect 32(1):42–45, 50 (in Chinese)

She N et al (2021) Reflections and Suggestions on China's Sponge city construction. Landsc Archit Front 9(4):82–91

Wang H et al (2015) Hydrologic control criteria framework in the United States and its referential significance to China. J Hydraulic Eng 46(11):1261–1271, 1279 (in Chinese)

Zhou Y et al (2021) Urban rain flood ecosystem design planning and feasibility study for the enrichment of smart cities. Sustainability 13(9):1–15

Zhu H et al (2017) Survey on residents' participation in Sponge city construction in Jiaxing city. CO-Oper Econ Sci 6(11):12–14 (in Chinese)

# Chapter 3
# Evolution Process and Distribution Characteristics of Ground Displacement Induced by Shield Tunneling

**Lingfeng Meng, Hua Jiang, Yusheng Jiang, Jinxun Zhang, and Yong Lv**

**Abstract** With the improvement of ground deformation control induced by shield tunnel crossing construction, in-situ field monitoring was carried out considering the necessity of research on deep ground displacement. Also, field test data were analyzed compared with numerical simulation analysis. The results indicate that the evolution processes of settlement troughs at different depths are complicated. For the monitoring points over the tunnel central line, the phenomenon of obvious deformation delay and ground stratification occur especially for those close to the tunnel vault, and the vertical displacements exceed others' until the shield tailskin passes the monitoring section. The displacement decreases as the horizontal distance from the central line increases and the whole settlement occurs with no obvious layered displacement existing beyond the tunnel excavation range. During the process of shield tunnelling, the ground displacements induced by the stages of shield passing through and shield tailskin leaving the monitoring points occupy 80–90% of the final settlement. The study can provide certain guidance for the establishment of deformation control in shield close tunnelling construction.

## 3.1 Introduction

With the rapid development of urban metro tunnel construction, the cases of new shield tunnels passing through existing tunnels, underground buildings and structures are gradually increasing, which has gradually become the main problem restricting the spread of underground space. In order to reveal the influence of shield excavation on deep strata and put forward corresponding control measures, evolution

L. Meng (✉) · H. Jiang · Y. Jiang · Y. Lv
School of Mechanics and Civil Engineering, China University of Mining and Technology (Beijing), Beijing 100083, China
e-mail: zqt1900605106g@student.cumtb.edu.cn

J. Zhang
Beijing Urban Rail Transit Construction Engineering Co., Ltd, Beijing 100088, China



P. Mendonça et al. (eds.), *Proceedings of 2023 International Conference on Green Building*, Environmental Science and Engineering,
https://doi.org/10.1007/978-3-031-43478-5_3

process and distribution characteristics of ground displacement should be studied. Many researchers have carried out detailed investigation into the characteristics of tunnelling-induced ground displacement (Zhang et al. 2017a, 2003; Mair et al. 1993; Han et al. 2007; Zhou and Jialiu 2002; Li et al. 2017). Based on theoretical derivation and numerical simulation, many prediction models or formulas of ground displacement have been proposed (Peck 1969; Attewell et al. 1978; Jiang et al. 2004), and some have been studied through field measurement (Rowe and Lee 1992; Loganathan and Poulos 1998; Xiongyu et al. 2016). However, due to the limitations of existing monitoring technology and numerical simulation methods, few researches on the distribution characteristics and development process of deep ground displacement caused by shield construction were carried out, which is necessary to be studied considering the effect of shield crossing on nearby pipelines, existing tunnels and other structures.

In this paper, based on the monitoring data acquired in a shield project of Beijing Subway, the development process and spatial distribution characteristics of ground displacement in the whole process of tunnel excavation are studied through on-site monitoring, aiming to provide guidance for the proposal of tunnelling-induced deformation control measures and control standards of adjacent buildings and structures.

## 3.2 Engineering Background

In order to reveal the evolution process and distribution characteristics of ground displacement caused by shield construction, the analysis of in-situ settlement monitoring of a shield tunnelling project of Beijing Subway is conducted, and the detail is referred in reference (Zhang et al. 2017b). The site is located in the plot to be developed, which is surrounded by walls and less affected by external factors. The soil strata of the site are relatively uniform, mainly composed of silt and silty clay. The monitoring points are arranged in five different depths, that is, the depths of 1.3, 4.3, 7.8, 9.3 m (i.e., the distance from the tunnel vault is 9, 6, 2.5, 1 in turn). The buried depth of tunnel vault is 10.3 m. Among them, Points a10 to a50 are the surface points, and points a11 to a51, a12 to a52, a13 to a53 and a14 to a54 are subsurface points (Zhang et al. 2017a).

Based on the requirements of field monitoring and the site conditions, this monitoring selects the technique presented in reference (Zhang et al. 2017a) applied in the field test, which combined anchor platform, static level and single-point extensometer together to form the comprehensive displacement monitoring system for more precise monitoring.

## 3.3 Analysis of In-Situ Monitoring Results

### *3.3.1 Displacement Distribution Characteristics of Different Depths*

According to reference (Zhang et al. 2017a), the monitoring data are actually the displacements of half the strata due to the monitoring points arranged between two tunnels, i.e., one side of the right tunnel axis. As the stratum distribution of the whole site is uniform, the displacement of the strata on both sides of the tunnel axis is considered as the same, and the settlement curve can be symmetrical with respect to the tunnel axis. Based on this assumption, the settlement curves are drawn as shown in Fig. 3.1. It indicates that the distribution characteristics of final displacements at different depths caused by shield excavation are similar, all of which present normal distribution settlement troughs. The displacements of the monitoring points directly above the tunnel vault shows certain stratification, which is more obvious when closer to the tunnel vault. Also, the displacements gradually decrease along each side of the tunnel axis, and the excavation contour is the boundary point of layered displacements, i.e., the displacements outside the excavation range occur as a whole without obvious ground stratification.

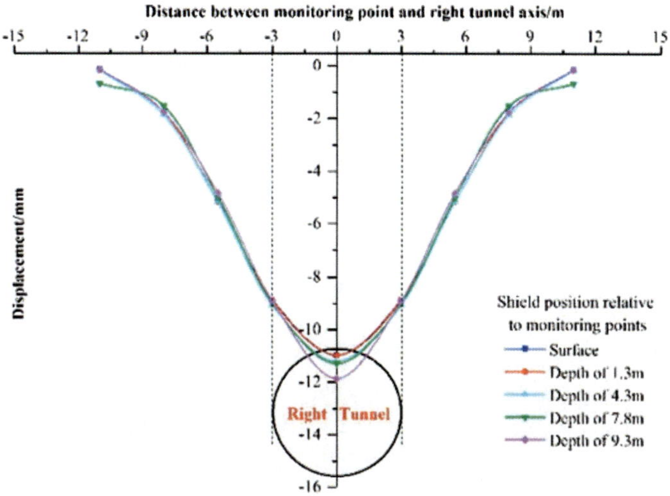


**Fig. 3.1** On-site monitored settlement troughs at different depths

### 3.3.2 Evolution Process of Ground Displacement at Different Depths

Figure 3.1 shows the final settlement troughs at the depths of 0 m (i.e., the surface), 1.3, 4.3, 7.8 and 9.3 m. Figures 3.2 and 3.3 present the evolution process of settlement troughs at depths of 0 and 7.8 m during the stages of shield reaching (−24.6, 15, 0 m away from the monitoring points), passing through (0.6 m), leaving (9 m), far away from (21 m) the monitoring points and long-term stability (65.4 m). The figures indicate that the evolution processes are nearly the same. Before the cutterhead reaches the monitoring points, the ground displacements are not obvious, and then the settlement troughs occur initially when the cutterhead passes through the points. Due to the extrusion effect on soil by the cutterhead, the bottoms present a large rebound, while the settlements on both sides of the bottoms continue to increase, thus causing double peak phenomenon of the settlement troughs. After the shield tailskin leaving, the displacements increase rapidly, the points directly above the tunnel vault subside fastest and still show rebound phenomenon but weakened, and the displacements are still smaller than those of the points on both sides of the axis. When the shield is far away from the points, the rebound of the points directly above the tunnel vault disappears, and the shapes of the settlement troughs are close to the characteristics of normal distribution. When the displacements tend to be stable, the settlement trough and the maximum settlement on each side remain unchanged.

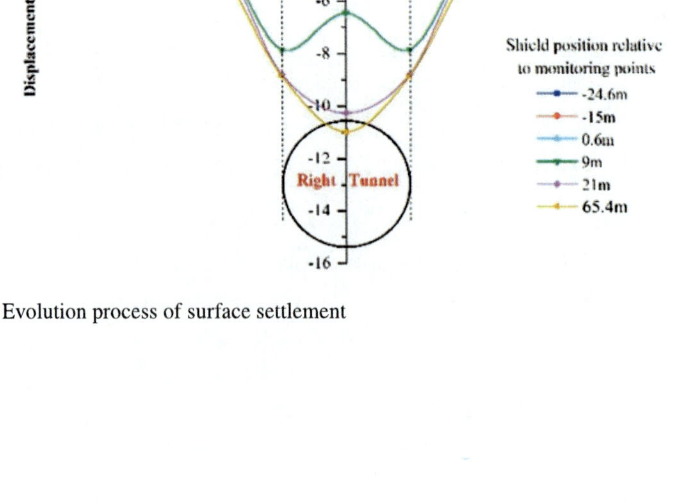


Fig. 3.2 Evolution process of surface settlement

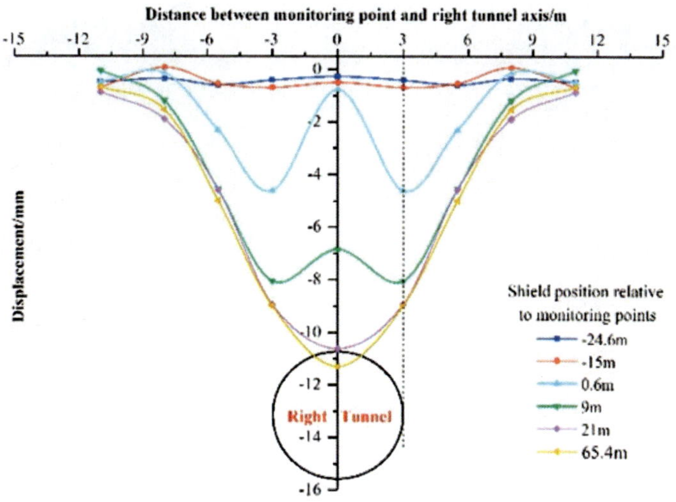


**Fig. 3.3** Evolution process of settlement at depth of 7.8 m

### *3.3.3 Ground Displacement Proportion at Different Stages of Shield Construction*

The shapes of settlement trough are changing dynamically before and after shield passing through the monitoring points, and the proportions of ground displacements at different stages are also different. For the convenience of research, the positions of shield relative to monitoring points are divided into four stages: prior to cutterhead arriving (0.6 m), shield passing (0.6–9 m), shield tailskin leaving (9–21 m), and the long-term settlement (21–65.4 m). The proportions of the increments of ground displacements at the surface and depth of 7.8 m are calculated in Fig. 3.4. It indicates that the characteristics of the displacement proportion at different depths are approximately the same. In the stage of prior to cutterhead arriving, the displacements account for a small proportion, about 6.0–11.3% of the total. When shield passes through the monitoring section, the displacements account for the largest proportion about 47.0–53.8% of the total, which is the main stage of displacement development. When shield tailskin leaves, the proportion is relatively large and accounts for 33.4–36.8%, which is the secondary stage of displacement development. In the long-term settlement stage, the proportion is the smallest and only accounts for 4.8–6.8%.

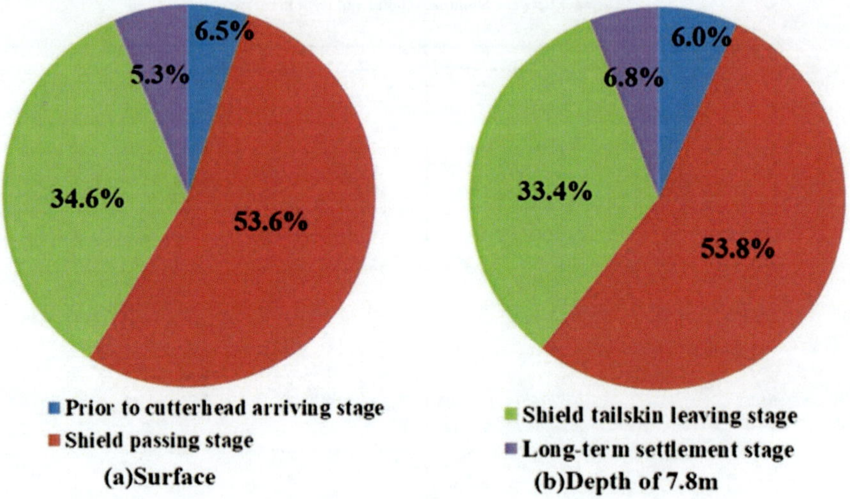


**Fig. 3.4** Monitoring analysis of displacement proportion at different depths

## 3.4 Numerical Simulation Verification and Comparative Analysis

### *3.4.1 Model Establishment and Monitoring Point Layout*

The ABAQUS numerical modelling is carried out and the geological conditions of monitoring section A are applied in the simulation. The model size is 70 m × 90 m × 30 m (width × length × height). Mohr Coulomb constitutive model is adapted for the soil material. The excavation diameter is 6.28 m, with the outer diameter of segment 6 m and the inner diameter 5.4 m. The gap between segment and soil is filled by synchronous grouting with 0.09 m thick ring. In order to simplify the calculation and ignore the hardening process of slurry, the final status of slurry is directly considered, and liner elastic material is selected for shield, segment and grouting layer due to the practical condition. The monitoring points are arranged on the surface and inside the stratum. The buried depths of the five horizontal monitoring lines from the surface to the tunnel vault are 0, 1.3, 4.3, 7.8 and 9.3 m, respectively, which is consistent with the practical layout, as is shown in Fig. 3.5. The soil parameters of different layers in the model are selected according to the geological survey report.

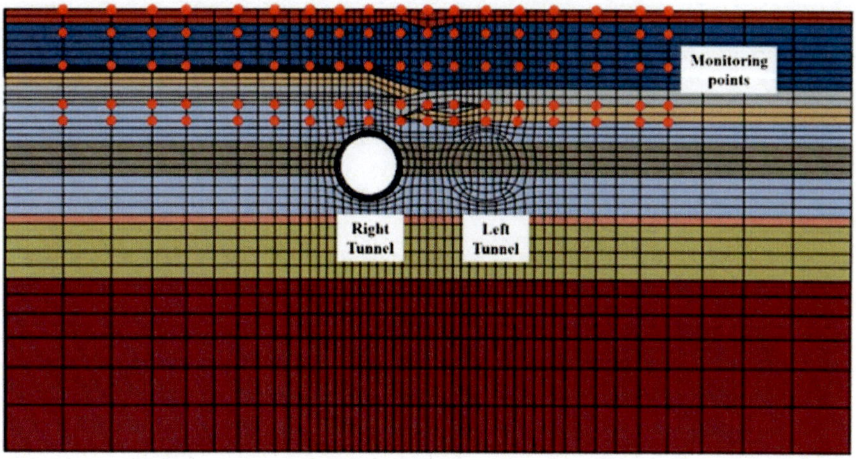


Fig. 3.5 Monitoring point layout of model

### 3.4.2 Simulation Results of Ground Displacement Evolution at Different Depths

The final displacements and evolution process of settlement troughs at different depths are analyzed in Figs. 3.6 and 3.7. Through comparison, the numerical evolution processes are basically consistent with in-situ monitoring results, while the numerical simulation cannot accurately reveal the soil heave effect at the initial stage of displacement evolution as well as the deformation lag phenomenon of the stratum directly above the tunnel axis. This is mainly due to the difficulty in simulating the changes of field tunnelling parameters such as thrust and earth pressure, while the complexity of settlement formation process is ignored.

Figure 3.8 indicates that the ground displacements generally present obvious normal distribution characteristics. With the increase of burial depth, the maximum displacement increases, and the layered difference within the range of tunnel excavation is obvious, which is consistent with the features revealed by in-situ monitoring. The numerical simulation results reveal clearly the phenomenon of displacement stratification, that is, settlement troughs with small depths are wide and shallow, while with large depths are narrow and deep.

### 3.4.3 Simulation Analysis of Displacement Proportion in Each Stage of Shield Construction

The simulation results are analyzed according to the four stages mentioned above in Fig. 3.9. Compared with the in-situ monitoring data, the simulation results indicate

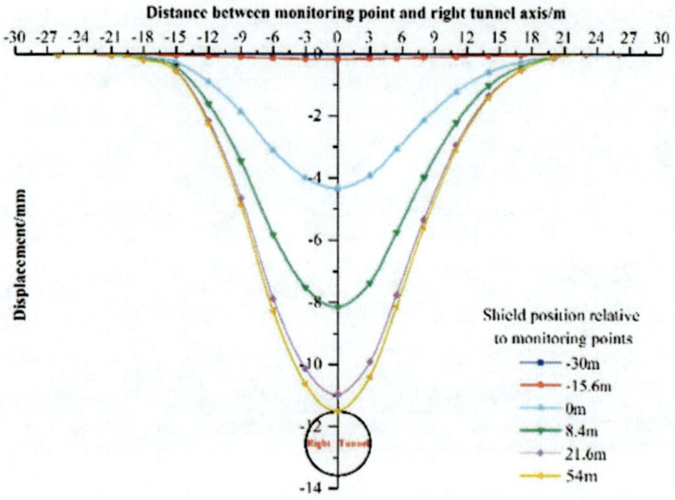


**Fig. 3.6** Numerical evolution process of surface settlement

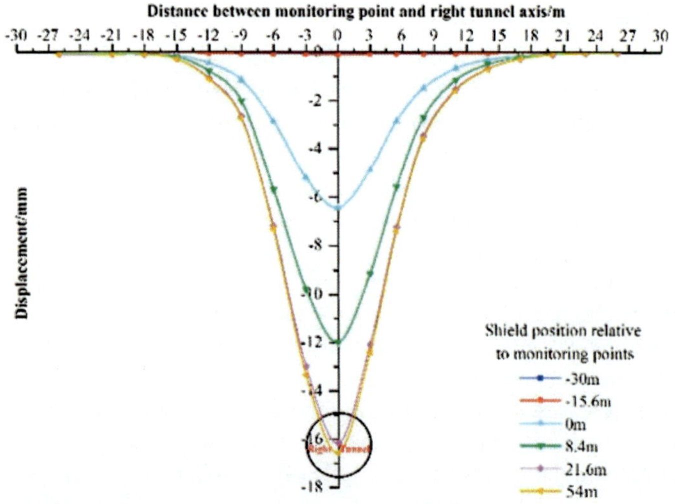


**Fig. 3.7** Numerical evolution process of subsurface at depth of 7.8 m

that the displacement proportions at different depths in each stage are closer, which still occupy a large part in the stages of shield passing and shield tailskin leaving (55.4–62.1%). Also, the proportion in the stage of shield passing is larger than that of shield tailskin leaving. The proportion of the long-term settlement stage is the smallest, consistent with the monitoring results. Nevertheless, the simulation results indicate that the displacement increment has a large proportion before cutterhead arriving due to the selection of numerical model parameters and the simulation of

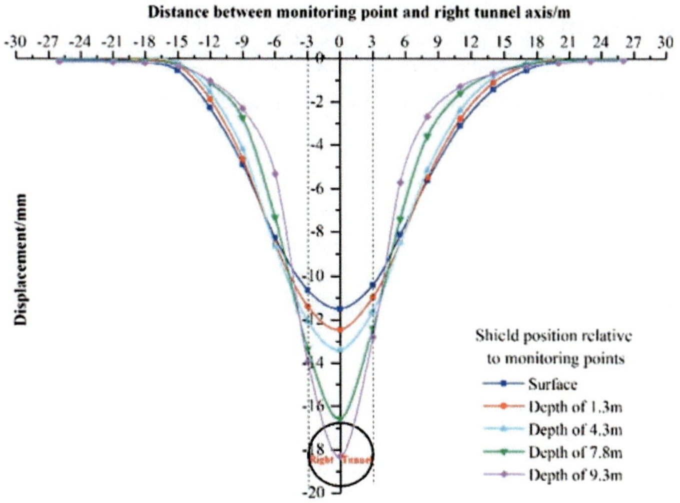


**Fig. 3.8** Settlement trough curves with different depths

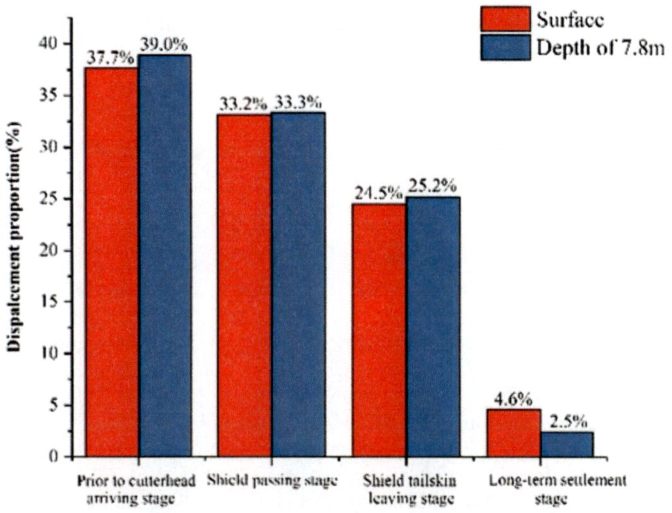


**Fig. 3.9** Numerical analysis of displacement proportion of each stage at different depths

excavation process, which can be regarded as a reference for field monitoring of shield construction.

## 3.5 Conclusions

Based on the comparative analysis of in-situ displacement monitoring data in shield tunnelling project and numerical simulation results, main conclusions are drawn as follows:

(1) The evolution processes of settlement troughs at different depths are complex during shield tunnelling, and the phenomenon of double peak effect and displacement stratification is obvious within the tunnelling influence range. Before the shield tail leaves away from the monitoring section, the point directly above the tunnel vault subsides later than other points.
(2) The distribution characteristics of the final displacements at different depths induced by shield excavation are similar, all presenting normal distribution. The points above the tunnel vault present displacement difference. The stratification phenomenon is more obvious when closer to the vault. The strata outside the tunnel excavation range present overall settlement without obvious stratification.
(3) In the process of shield excavation, the displacements in the stages of shield passing and tailskin leaving away from the monitoring section account for 80–90% of the total. In order to realize the micro control of surface subsidence and soil deformation induced by shield passing through existing buildings and structures, the construction control measures of the above two stages must be done well.

## References

Attewell PB, Glossop NH, Farmer IW (1978) Ground deformations caused by tunnelling in soil. Ground Eng 15(8):32–41

Han X, Li N, Standing JR (2007) An adaptability study of Gaussian equation applied to predicting ground settlements induced by tunneling in China. Rock Soil Mech 28(1):23–28, 35

Jiang XL, Zhao ZM, Li Y (2004) Analysis and calculation of surface and subsurface settlement trough profiles due to tunneling. Rock Soil Mech 25(10):1542–1544

Li T, Li D, Fan K et al (2017) The study of pile foundation displacement caused by adjacent excavation of underground super large space. J Min Sci Technol 2(6):529–538

Loganathan N, Poulos HG (1998) Analytical prediction for tunneling-induced ground movement in clays. J Geotech Geoenviron 124(9):846–856

Mair RJ, Taylor RN, Bracegirdle A (1993) Subsurface settlement profiles above tunnel in clays. Geotechnique 43(2):315–320

Peck RB (1969) Deep excavations and tunnelling in soft ground. In: Proceedings of 7th international conference on soil mechanics foundation engineering. Mexico City, State of the Art Volume, pp 225–290

Rowe RK, Lee KM (1992) An evaluation of simplified techniques for estimating three-dimensional undrained ground movements due to tunneling in soft soil. Can Geotech J 29:39–52

Xiongyu H, Yan Q, He C et al (2016) Study on the disturbance and excavation face failure feature of granular mixtures stratum due to EPB shield tunnelling. Chin J Rock Mech Eng 35(8):1618–1627

Zhang YP, Yu YN, Zhang TQ (2003) A new prediction model with time-dependent parameters used for settlement prediction. Chin Civil Eng J 36(12):83–86

Zhang J, Jiang H, Jiang Y et al (2017a) Measurement technology for strata displacement at different depths induced by shield-driven tunnelling. Modern Tunnel Technol 54(4):123–130
Zhang J, Jiang H, Cheng J et al (2017b) Spatial distribution prediction method of ground displacement induced by shield tunnelling. J Railway Eng Soc 34(11):88–94
Zhou X, Pu J (2002) Centrifuge model test on ground settlement induced by tunneling in sandy soil. Rock Soil Mech 23(5):559–563

# Chapter 4
# Based on the Causes of the Opening and Characteristic Space of Different Types of Universities

**Chenming Yao and Zhuomin Lin**

**Abstract** Nowadays, the contradiction between the shortage of land in the central area of the city and the rapid expansion and development of colleges and universities is becoming increasingly prominent, and the intersection of colleges and universities and cities is becoming closer. Thus, the issue of whether colleges and universities should exist as an independent individual or should be open and integrated with the city has entered the public view. Therefore, this paper focuses on the development process of colleges and universities and relevant background theories, studies the influencing factors and causes of different open mode universities, and deeply discusses their relationship with cities. Taking South China University of technology and Hunan University as comparative cases, this paper makes a comparative analysis from the aspects of traffic characteristics, public elements and management measures, summarizes the advantages and disadvantages of a relatively closed campus and a relatively open campus, and comprehensively extracts the advantages to improve the level and dimension of the integration of University and urban space, resources and personnel, and from the campus boundary, transportation, industry, public resources After combing and summarizing the five perspectives of security management, the corresponding optimization strategies are put forward to create a new universal open campus mode, and provide social, public and spatial development ideas for colleges and Universities under reconstruction or new construction.

**Keywords** Open campus · University boundary form · Traffic characteristics · Public elements · Management measures · User feelings

C. Yao (✉) · Z. Lin (✉)
Faculty of Architecture, South China University of Technology, Guangzhou, China
e-mail: 776369173@qq.com

Z. Lin
e-mail: winnie980131@163.com



P. Mendonça et al. (eds.), *Proceedings of 2023 International Conference on Green Building*, Environmental Science and Engineering,
https://doi.org/10.1007/978-3-031-43478-5_4

## 4.1 Foreword

With the process of urbanization, the construction and transformation of universities have also developed to a key stage (Zeng and Xu 2021). Looking at the site selection and construction of domestic universities, the universities built by relying on the development effect of urban prosperity are the mainstream, and most of the universities are located in the urban public space. Due to the comprehensive use of the characteristics of developed urban economy and convenient transportation, and in the subsequent development process, the scale of colleges and universities has expanded, resulting in the loss of university development space in the central city, so the model of university town came into being. Universities worldwide now encounter far greater challenges, and are subjected to an unprecedented level of external scrutiny. The change in governance ideology in the higher education sector has altered the way in which universities are managed (Yang 2004). At the same time, the form of colleges and universities also under the influence of the closed unit compound system, the campus boundary space using walls or railings entity boundary form, isolation of the campus space, the reason, this conflicts with the growing traffic demand and urban land resources tension, and limit the colleges and universities and the outside world and contact the ability of resource circulation is closely related. Based on this phenomenon, and due to the continuous expansion of contradictions, universities began to try to integrate with the urban space, embedded as a part of the city, forming a win–win situation in the continuous cooperation and interaction with the city. And this type of colleges and universities generally through fuzzy boundary space to contact city and society, and different colleges and universities in the actual construction and management mode and keep closed schools in many aspects, therefore, this paper focuses on the city center of different open mode of colleges and universities cases, especially the processing of the social space and campus internal space and different boundary form, get the future university space design reference, promote the common development of cities and colleges and universities.

## 4.2 Theoretical Research

### *4.2.1 Open Campus Definition*

At present, the definition of college campus opening is mainly reflected in the following two aspects: physical space opening and hidden resources opening. The former mainly refers to the establishment of multi-level sharing space and activities to meet and support the needs of university personnel and social personnel, such three-dimensional contains rich participation space can enrich campus sharing framework, (Peker and Ataöv 2020) on the premise of ensuring open space continuity, create multi-level public space/communication space, improve the liquidity and variability of campus space, and through continuous order of open space, to enhance the campus space of urban openness and accommodation. The latter mainly refers to the university academic/scientific research/culture/industry and other non-explicit resources sharing, its not only to serve the campus of the educated, more open to the school at the same time, realize resource sharing, the get the maximum equal use of campus resources, integrate education resources and scientific research resources, form the corresponding framework system, can make the better sharing resources, make university resources are no longer in campus insiders as the main body of the resources.

As the University campus becomes ever more digitally connected, it enables new forms of collaboration and communication (Alicea 2021). However, based on the current campus environment security and management mode, many colleges and universities in both hope toward the open development at the same time, and have to face the actual only half-open or specific form of open, to restrict liquidity (such as China workers), but there are (such as lake) universities successfully completely open in a sense, therefore, based on the interpretation of open campus, the different degree of openness in the corresponding categories, to make specific explanation.

### *4.2.2 Different Boundaries*

In campus planning, boundary space is the main space to communicate with internal and external space. Just as Aldo Fan Ike believes that "a building is a city, and a city is a building" (Zhang 2005), it not only plays the role of dividing the texture space of the city, but also plays the role of a bridge between society and universities. The boundary morphology of campus which mainly consists of space, function, transportation etc. is the connecting and transiting space between the campus and the city. The reasonability of its form not only affects the improvement of urban function and environment, but also connects with the external image of the campus, furthermore, it has a great impact on psychology and behavior of people, especially university teachers and students, who make use of the space (Xu 2014). Therefore, the boundary space of the two cases is divided into three types according to different forms, namely, the boundary space based on the campus space structure, that dominated by soft partition such as landscape greening, and the boundary space dominated by comprehensive auxiliary functions.

#### 4.2.2.1 The Boundary Space is Based on the Campus Space Structure

This kind of space is the use of the overall campus planning and layout to surround the space, and the edge of the space is the boundary of this kind of space. In fact, such space does not belong to the category of boundary, but produces different boundary effects under the influence of core functions. Therefore, different internal layout methods will affect the existence or not of the boundary and the feeling of the boundary space. (Figure for language, for example).

#### 4.2.2.2 The Boundary Space is Dominated by the Soft Partition Such as Landscape Greening

This type of space is generally in the actual expansion of campus construction, and the use of urban space, lead to space use, soft partition and use can effectively reuse resources, for example, to landscape greening and square space for flexible design, improve the permeability and continuity of space. (Figure for language, for example).

#### 4.2.2.3 A Boundary Space Dominated by Accessory Functions

Due to the intensive use of urban land resources, the campus boundary and its surrounding space can be shared to a certain extent, thus creating the dominant boundary space of the type of space, through the commercial space with intermediary nature, to play the role of outward extension and inward guidance, and provide services for the people on both sides of the market. Due to its clear auxiliary positioning, this kind of space surrounds the campus with special determination techniques and ensures its corresponding position.

According to the summary of the above types, it can be seen that different types of boundary space will affect the openness of the campus to some extent due to the disunity of the traditional boundary, and the campus will also transform from the state of separation from the city to the result of positive integration and symbiosis.

## 4.3 Comparison Between the Two Schools

The following comparative analysis data are mainly carried out through the literature review method, field survey method, questionnaire survey method and interview method.

### *4.3.1 Overview Comparison*

This paper selects Wushan Campus of South China University of Technology (hereinafter referred to as SCUT) as the main research object to represent the management mode of merger and reorganized universities. It is located in the center of Tianhe District, Guangzhou, and the radiation of surrounding universities and the support of corresponding educational resources have formed a group of universities in the area dominated by SCUT. It covers an area of more than 2.94 million square meters, and the whole campus is disrupted by the ups and downs of the mountains, and the road generally has a slope/Angle, making SCUT form a unique texture integrated with the mountain terrain.

Through the study of historical texture evolution process analysis, found that SCUT under the premise of compound development, early to adapt to the terrain and planning to lay the layout of the bell plane, medium reorganization merger conform to demand to adjust the campus pattern, late and due to the development of urbanization construction density strengthening and conflict and urban land use, therefore, in the long development and reproduction change, formed the pattern of the existing enveloped.

And Hunan university south campus (hereinafter referred to as HNU) is representing the basis of open university, it is located in the Yuelu mountain, surrounded by natural landscape, natural terrain development and related requirements and reconstruction and the city under the integration formed no strict pattern of walls and school, and the city without obvious dividing line, it covers an area of about 2 million square meters, campus through the work and study area and living area, form its unique open form.

For its development cause, the uniqueness of HNU mainly inherited the academy school tenet and education thought, due to the authoritative and solemn academy system in history (Wan 2015), makes the early HNU planning centered on Yuelu academy, and because of modern education progress, yuelu academy with Yuelu mountain evolved into a HNU spiritual symbol and historical witness, become an open historical scenic area, and the campus space in the process of scenic area gradually blend formed the joint open to the society.

Within a long course of the history of higher education institutions, they have expressed various types of integration or separation regarding their adjacent urban space (Fard et al. 2022). Through comparison, it can be seen that relatively closed universities (such as SCUT) are generally based on the early terrain planning and focusing on the campus functional spatial layout. Under the protection strategy based on the overall layout system in response to the constraints of urban development, a pattern with certain boundary subjects is formed. However, the existence of a special fully open university (such as HNU) provides an internal campus based on its own group functional layout (mainly in Yuelu), but also constantly develops simultaneously with the drastic changes of the external social environment, leading to the continuous adjustment of its own form, naturally ignoring the boundary and open.

### 4.3.2 Comparison of Traffic Characteristics

SCUT surrounding road network structure is complex, hierarchical, through the large transportation hub, public transportation and walking system, on the one hand, for the personnel inside and outside the travel selectivity, on the other hand, because the campus block the city group, lead to campus external vehicles had to detour, increase the campus surrounding traffic pressure, and in order to ensure the important road unobstructed, the campus is divided into the north and south two area, and it also brought the campus north and south area connecting traffic flow, traffic mixed, traffic safety hidden trouble. And SCUT introverted traffic also has a great impact on the transportation mode of teachers and students (Qin 2018), Making electric cars and walking the main way of travel, It also shows that the lack of some categories of transportation will affect the study and life in school, At the same time, the internal traffic planning of the campus itself does not consider the traffic and guidance of vehicles outside the campus to a large extent, The setting of the school gate and the relatively closed management limits the accessibility, Therefore, schools with open campus policies had higher mean walkability scores than closed campus schools. Meanwhile, district and school policy decision-makers may take into account the walkability around the school when considering open campus policies (Budd et al. 2020). The relatively closed campus traffic in the relationship between the city and the traffic flow is weak and it is difficult to provide traffic guidance for the city, While SCUT was supposed to undertake more urban functions because of the complexity of its surrounding traffic, But they are all affected by the occlusive campuses.

However, the road network near HNU is relatively simple, and it is adjacent to the main roads in fewer cities, and the large transportation hubs are far away. Therefore, it is less affected by the traffic burden in the morning and evening rush hours.At the same time, the opening of the campus internal traffic and scenic spot traffic, combined with the campus bear most of the transit traffic, social vehicles through the campus to avoid a wide detour, but it also brought the corresponding negative impact, social vehicles mixed, increase the campus safety hidden trouble, with the interoperability of public transportation, provide convenience for teachers and students to commute at the same time, will also bring a large number of tourists from Yuelu mountain to the campus. For example, because the subway station is located inside the campus, it provides more selectivity for teachers and students to commute, and causes an increase in the overlap of school people and social people. Therefore, although the relatively open campus of campus road transit vehicle control put forward strict requirements to deal with the open traffic safety hazard, but it promotes the extroversion of the mode of traffic and in the urban traffic and on the integration of urban traffic degree has more significant advantages, can promote the maximum use of traffic resources.

### *4.3.3 Comparison of Public Elements*

Open form of campus boundary removed walls and gates to achieve resource sharing and functional complementation (Xu 2014). From the perspective of the distribution of scientific research industries, universities are both significant knowledge enterprises and the suppliers of the human and intellectual capital on which the knowledge-based economy depends (Benneworth et al. 2010), the research centers of SCUT and its surrounding are mainly gathered in the campus, while there are few business incubators and related industrial science and technology parks radiating to the surrounding campus, and they are far away from the single function in other scientific research industrial chains with the same functions in the city. As for business perspective (Table 4.1), in the service of SCUT business mainly distributed in the campus periphery and dot focus on the campus entrance and exit (Ning 2018), business type is mainly suitable for college students, for the purpose of meeting the student group, so with the campus crowd flow seasonal fluctuations, business inevitably appeared in the summer season, business performance and crowd flow are relatively low, this performance in the low fusion with urban schools (such as SCUT) is widespread and significant. In terms of public facilities (Table 4.3), most of the campus facilities are evenly dispersed in the functional groups of the campus, but there are a small number of facilities that lack regularity, and the total number of facilities still cannot meet the per capita needs. Moreover, due to the closed management of the campus, the lack of total facilities makes the proportion of public facilities unbalanced, and the publicity of the campus facilities is weak, which cannot flexibly meet the people inside and outside the school at all levels. However, the trough of seasonal fluctuation and the use of facilities are excessive, so it is particularly important to coordinate the public nature and meet the seasonal high and low peaks.

On the other hand, HNU, compared with SCUT, has established public facilities related to the scientific research industry in the integration area of campus and city, and integrated the academic research advantages of the surrounding universities in the radiation area, so as to realize the maximum use of campus knowledge and improve the efficiency of the downstream applied science industry. At the same time, the campus commerce mainly presents a unique linear distribution form along the campus boundary and the city integration place, and the folds along the main roads of the campus. (Table 4.2) Moreover, the total amount and permeability of commercial facilities are large, which improves the convenience of teachers and students to a greater extent. Due to the high cross-integration degree of the social personnel and the campus people flow, the seasonal volatility of the open campus people flow will

**Table 4.1** Commercial facilities around SCUT (This chart is based on figures from Qin Yanan, Research on the edge space optimization strategy of Wushan campus of South China University of Technology based on interactive mechanism, 2018)

| Type of business | Contents | Area per capita (per/ $m^2$) |
|---|---|---|
| Meet the basic needs of the university | Snack bars, convenience stores, restaurants, milk tea shops, hotels, florists, digital, pharmacies, clothing stores, optical stores | 0.44 |
| Meet the high-level needs of universities | Bookstore, café, pantry, gym<br>Renderings companies | 0.12 |
| Meet university-specific needs<br>Generated by university | Renderings companies, graphic companies, model companies, training institutions | 0.06 |
| Generated by university resources | Design companies, cultural media companies, industrial companies, Internet companies, technology companies | 0.07 |
| Total area of commercial facilities | 31847.56 $m^2$ | 0.69 |

**Table 4.2** Commercial facilities around HNU which is based on figures from field investigations by authors

| Type of business | Contents | Area per capita (per/ $m^2$) |
|---|---|---|
| Meet the basic needs of the university | Snack bars, convenience stores, restaurants, milk tea shops, hotels, florists, digital, pharmacies, clothing stores, optical stores | 1.59 |
| Meet the high-level needs of universities | Bookstore, café, pantry, gym<br>Renderings companies | 0.25 |
| Meet university-specific needs<br>Generated by university | Renderings companies, graphic companies, model companies, training institutions | 0.09 |
| Generated by university resources | Design companies, cultural media companies, industrial companies, Internet companies, technology companies | 0.08 |
| Total area of commercial facilities | 51264.2 $m^2$ | 2.00 |

not have a significant impact on the business performance. In addition to the life service facilities that can be evenly distributed throughout the campus, the other facilities (Table 4.3) are mostly centered on the campus boundary, and the per capita facility area is introduced into the urban municipal facilities due to the open campus, which is more sufficient compared with SCUT. Therefore, the distribution area of richness and advantages can improve the use efficiency of different facilities, and it can be flexibly handled according to the needs of different groups.

### *4.3.4 Comparison of Management Measures*

In our country, access control management is one of the important means to ensure the safety of teachers and students, and for such as HNU—an open campus, external access has greater flexibility, for SCUT such kind of universities, limit the access of outsiders is one of the elements to reduce its openness. For any campus at home and abroad, the strict management of dormitories is an important guarantee to determine the personal safety of students on campus, such as the access control management of SCUT and HNU, the real-name reservation of Taiwan University and Oxford University, and the management of limiting the number of visitors. For SCUT, the setting of external campus walls, clear boundary lines and strict access control management, which limit the entry of external personnel, so that the open space inside the campus does not need to take a greater intensity of management. Therefore, the sharing of campus space with the city brings challenges to the management measures, and makes the use of campus space more extroverted. For example, HNU is divided by urban roads, with a single building as the management unit, the space is open to the

**Table 4.3** Comparison of information on interactive elements of universities

| Type of facilities | Contents (SCUT) | Layout features | Contents (HUN) | Layout features |
|---|---|---|---|---|
| Landscape environment | Landscape lake, green space | It concentrated in the North and Central, and most is close to the center of the campus | Landscape lake, green space | The landscape area is relatively small, it is centrally arranged in the public teaching area |
| Sports facilities | Playgrounds, gymnasiums, etc. | Evenly distributed | Playgrounds, gymnasiums, etc. | Distribution on both sides |
| Educational facilities | Municipal public education service facilities such as kindergartens, primary schools affiliated to SCUT | Near the boundary but scattered | HNU schools affiliated Middle School and other municipal public education service facilities | Near-boundary centralized layout |
| Health facilities | University Hospital (listed as a community health service center) | Near campus boundaries | University Hospital | Near campus boundaries |
| Cultural and leisure facilities | Cultural and Sports Center, Faculty Activity Center, cafeterias, stores, etc. | Small quantities; Near city roads | Yuelu Academy, Inferiority Pavilion, etc. | Concentrated near the Yuelu Academy |
| Supporting facilities for life services | On-campus cafeterias, convenience stores, etc. | Evenly distributed | On-campus cafeterias, convenience stores, etc. | Equally distributed |
| Scientific research and production facilities | University Science Park, school-run enterprises, Architectural Design and Research Institute, etc. | Arranged near campus boundaries | International Supercomputing Center, Physics Experimental Middle School, etc. | Near to the edge of the campus layout |
| Total area of shared service facilities | 123000 $m^2$ | 2.66 $m^2$/person | 93000 $m^2$ | 3.64 $m^2$/person |

outside world, and the public at different peak use, management has no limit, fully elaborated the specific practice of open campus to adopt more refined management measures. In addition, the overseas campus open space is also fully open, through the formation of channels in multiple sections to limit the use, or the rich and diverse management modes of peripheral opening and internal reservation access are also worth learning from.

### *4.3.5 Users Feeling Comparison*

Since SCUT mainly has hard boundaries such as walls, slope protection, railings, by investigating campus space users, most people have a strong sense of belonging to campus life (Peker and Ataöv 2020), and only a few people think that the security of campus boundaries is poor. Through data comparison (Tables 4.4, 4.5), it is not difficult to see that the campus sense of belonging and boundary security has certain corresponding connection, and for open university such as HNU, rigid boundary / neutral boundary and natural boundary distribution has certain order, at the same time, campus space security and its boundary form has corresponding connection. In addition, the two schools have a good feeling of the use of learning space, and HNU is better. But there are some differences, SCUT are mainly because the space is quiet and and the natural environment is comfortable, and HNU has high accessibility of Yuelu Mountain. On the contrary, for people with poor feelings of space use, SCUT is mainly due to poor accessibility and lack of public space for academic communication, while HNU comes from the mixed noise of people and the disturbed teaching environment caused by the campus opening. Therefore, it can be concluded that due to the closed management, the campus learning space is relatively quiet and comfortable, and the learning experience is relatively good. However, due to the limited choice of paths, the commuting cost is relatively large. Therefore, the above problems can be solved to a certain extent through the transformation of the open campus management mode.

Through the questionnaire, (Tables 4.6, 4.7) comparing the campus students for the campus surrounding cities, it can be seen that the teachers and students in HNU have better understanding of the city, especially reflected in urban business and culture, so it can be seen that because of its close connection with the city, students' awareness of cities are higher and wider in open campus.

For the people outside the school, most residents, whether HNU or SCUT, have a high understanding of the campus. The surrounding residents of SCUT hope to use the campus infrastructure to a greater extent and have deeper communication with students. However, the management mode of SCUT has caused obstacles to the communication mode inside and outside the city. As for the HNU, due to the lack of personnel difference inside and outside the school, its needs are more inclined to have targeted solutions under the unified management.

**Table 4.4** Relevant metrics about campus familiarity based on more 500 people's questionnaires

| Judgement | | Sense of belong (%) | Living conditions (%) | Safety on border (%) | Judgement |
|---|---|---|---|---|---|
| SCUT | Well | 12.5 | 20 | 47.5 | Good |
| | Good | 27.5 | 20 | | |
| | Just so so | 60 | 58 | 52.5 | Just so so |
| | Bad | 5 | 2 | 2.5 | Bad |
| HNU | Well | 20.97 | 27.42 | 14.52 | Good |
| | Good | 35.48 | 53.23 | 75.81 | Just so so |
| | Just so so | 38.71 | 19.35 | | |
| | Bad | 9.68 | 3.23 | 16.13 | Bad |

**Table 4.5** Relevant indicators of awareness on and off campus based on more 500 people's questionnaires

| Degree | | Knowledge of outside (%) | The inconvenience of interior (%) | Degree |
|---|---|---|---|---|
| SCUT | Know traffic | 67.5 | 52.5 | Traffic inconvenience |
| | Know commercial | 42.5 | 47.5 | Shopping inconvenience |
| | | | 25 | Activity inconvenience |
| | Know culture | 32.5 | 20 | Communication inconvenience |
| | No ideas | 40 | 20 | No inconvenience |
| | Others | 2.5 | 2.5 | Others |
| HNU | Know traffic | 69.35 | 45.16 | Traffic inconvenience |
| | Know commercial | 61.29 | 20.97 | Shopping inconvenience |
| | | | 29.03 | Activity inconvenience |
| | Know culture | 54.84 | 9.68 | Communication inconvenience |
| | No ideas | 19.35 | 32.26 | No inconvenience |
| | Others | 1.61 | 6.45 | Others |

**Table 4.6** Relevant indicators about campus feelings based on more 500 people's questionnaires

| Judgement | | Points | Percentage (%) | Percentage (%) | Points | Judgement |
|---|---|---|---|---|---|---|
| SCUT | Good | Quiet | 70.83 | 54.71 | Accessibility | Bad |
| | | Cosy | 66.67 | 41.18 | Communication | |
| | | Sense of safety | 33.33 | 23.53 | Sense of safety | |
| | | Accessibility | 33.33 | 17.65 | Uncomfortable | |
| | | Traffic | 20.83 | 11.76 | Noisy | |
| HNU | Good | Quiet | 58.70 | 57.89 | Quiet | Bad |
| | | Cosy | 54.35 | 47.37 | Cosy | |
| | | Sense of safety | 41.30 | 42.11 | Sense of safety | |
| | | Accessibility | 34.78 | 42.11 | Accessibility | |
| | | Traffic | 34.78 | 26.32 | Traffic | |

**Table 4.7** Relevant indicators about how people on campus feel about communication based on more 500 people's questionnaires, mainly focus on students and residents

| Judgement | | Degree of knowledge (%) | Judgement | Degree of communication with students (%) | Judgement | Influences of campus close (%) |
|---|---|---|---|---|---|---|
| SCUT residents | Know well | 44.44 | Always | 44.44 | Facilities | 33.33 |
| | Just so so | 44.44 | Frequent | 33.33 | Traffic | 11.11 |
| | | | Sometimes | 11.11 | Communications | 11.12 |
| | No idea | 11.12 | Never | 11.12 | No | 22.22 |
| | | | | | Others | 22.22 |
| HNU residents | Know well | 48 | Always | 28 | Facilities | 15 |
| | Just so so | 39 | Frequent | 44 | Traffic | 20 |
| | | | Sometimes | 20 | Communications | 35 |
| | No idea | 13 | Never | 4 | No | 17 |
| | | | | | Others | 13 |

## 4.4 Sum Up

Comprehensive the above comparative analysis, relatively open campus, can alleviate the development of rapid urbanization of urban land tension and the traffic pressure of campus surrounding cities, to a certain extent improve the travel convenience of teachers and students, at the same time open campus can improve the utilization rate of campus public resources, promote the development of related industries. For users, the more open campus increases the opportunities for mutual communication, enhances the communication degree, promotes two-way penetration, and for the university itself, it expands the publicity of the boundary, reduces the distance between the campus and the city, and makes the integration degree closer. In general, the open campus to city integration, industry, optimize urban transportation and public resources have greater positive impact, on the contrary, open campus in campus safety management and students campus space using the negative impact is bigger, mainly reflected in the social vehicles and foreign personnel mixed with safety hidden danger and security trust reduced. However, combined with actual cases, the disadvantages of openness can mostly be avoided from the management mode.

The traditional closed campus teaching mode as a form of education recognized by the public, in the growing process of urbanization, and urban demand conflict, a single land properties and low density development mode has not conform to today's resource allocation mode (Chen and Lei 2012), at the same time, even the relatively open campus, surrounded by walls, its only open and advocate the concept of compound development is contrary. Therefore, the internal and external resources need to get a greater degree of sharing, and sharing and private need to get targeted guarantee on management, the corresponding contradiction, cannot be from a single open or not open to solve, and need to comprehensively extract the advantages of colleges and universities, the personnel and resources to the same height, the urban resources, university resources, urban personnel, university personnel in four-dimensional contact.

## 4.5 Optimization Advice

Based on the above summary, drawing on the strengths of the different types of openness in universities and coming up with practical openness strategies is a topic that can meet the meaningfulness of today's social context. Concretely, this is achieved by the university collaborating with diverse partners and stakeholders to develop (Trencher et al. 2013). In response to the aspects analyzed above, targeted strategies are proposed, summarized in the following five points.

### *4.5.1 Boundary Opening Strategy*

Reducing the use of hard borders, using neutral and soft borders as the main form of borders, and creating spaces with practical functions or sequences around the periphery of the campus can liberate the traditional borders from their single function, and enhance the hierarchy and richness of border spaces at the combination of city and school, transforming the periphery from absolute division to active connection and integration. For spaces and clusters with special needs, such as teaching, dormitories and other spaces for private activities of students and teachers, hard boundaries are adopted to ensure their functional integrity.

### 4.5.2 Traffic Opening Strategy

The static available traffic spaces, such as car parks, are given over to the dynamic urban roads to maximize the efficiency of transit traffic on campus, but at the same time, to ensure that they do not encroach on the space of the campus itself, the urban roads are used to divide the clusters, establish a cluster-based management model, connect the clusters to the peripheral communities, establish a rich pedestrian system and create a shared slow walking system on and off campus. In addition, the application of municipal traffic rules to the interior of the campus reduces traffic safety hazards within the campus and mitigates the disadvantages of open traffic in terms of traffic management.

### 4.5.3 Industrial Opening Strategy

Universities should provide talents and technical resources for high-tech enterprises, and the development of enterprises feeds the campus. For off-campus, a mixed functional area can be created around the campus to attract investment and businesses from related industries, creating a complete and benign living community adapted to the campus. For on-campus space, businesses can be introduced and flexibly distributed to promote the development of commercial types within the university and the convenience of life for teachers and students on campus.

### 4.5.4 Public Resources Strategy

Akinsanmi suggests that people learn better in a challenging, safe, comfortable, social and enriched environment (Peker and Ataöv 2020). Open spaces (including newly developed spaces, border crossing spaces and surrounding communities) will be integrated and packaged with resources, while opening them to different groups of people to improve the efficiency of resource use and ensure the integrity of services to the users.

### *4.5.5 Security Management Opening Strategy*

The management of shared and private spaces will be differentiated, and the intensity of management will fluctuate according to demand at different times and seasons.

## 4.6 Peroration

With the closer relationship between universities and cities, higher interactivity and integration, universities located in important locations of cities need to embrace the city in a more open attitude, so as to promote the participation and interaction in culture, economy, personnel exchange, material exchange and other aspects. Therefore, the purpose of optimizing the design of the open mode of colleges and universities is to promote the symbiosis between colleges and universities and cities. Through the border opening, traffic integration, industry linkage, resource sharing, strengthen the safety of the optimization strategy of the management, the internal space will be constantly improved, form the open pattern of order, and on the city with more continuous, both solve the problem of urban land tension and open to college campus and urban environment are greatly helpful.

The process of interaction between universities and cities is now seeing academics—and not just universities—at the forefront of the search for contexts to transform research into spaces of social intervention and open campus, giving a transformative orientation to research (Busacca 2020).

University campuses are place-based large institutions which create a direct interaction with their surrounding urban setting. Universities are shaping and being shaped by their urban context (Fard et al. 2022). The more open and open future campus will be an important medium to promote the four-dimensional connection of city resources, university resources, city personnel and university personnel and make their integration, sharing and interaction. Therefore, the theoretical strategy of summarizing the universality covers the spatial design of universities in different urban locations, which lays a foundation for the subsequent combination of theory and practice to promote the integration of universities and cities.

## References

Alicea B (2021) Open-source campus. Figshare

Benneworth P, Charles D, Madanipour A (2010) Building localized interactions between universities and cities through university spatial development. Eur Plan Stud 18(10):1611–1629

Budd E, Liévanos R, Amidon B (2020) Open campus policies: how built, food, social, and organizational environments matter for oregon's public high school students' health. Int J Environ Res Public Health 17(2):469

Busacca M (2020) Academics are back in town: the city-university relationship in the field of social innovation. Soc Mutam Polit (firenze) 11(21):187

Chen X, Lei R (2012) Review and prospect of the evolution of campus forms in Chinese universities. Planner 28(02):86–93

Fard HR, Trisciuoglio M, Yüksel D (2022) The morphological side of university-city interaction. Appl Mech Mater 584–586(2014):30–234

Ning CAI (2018) Study on the changes of South China University of Technology based on historical map. Chin Foreign Arch 02:82–86

Peker E, Ataöv A (2020) Exploring the ways in which campus open space design influences students' learning experiences. Landsc Res 45(3):310–326

Qin Y (2018) Research on the edge space optimization strategy of Wushan campus of South China University of Technology based on interactive mechanism. South China University of Technology

Trencher G, Yarime M, Kharrazi A (2013) Co-creating sustainability: cross-sector university collaborations for driving sustainable urban transformations. J Clean Prod 50:40–55

Wan L (2015) Study on the spatial morphology evolution of Hunan University campus. Hunan University

Xu Y (2014) Research on boundary morphology of campus-taking the Yangtze River delta region as an example. Appl Mech Mater 584–586:230

Yang R (2004) Openness and reform as dynamics for development. High Educ 47(4):473–500

Zeng F, Xu H (2021) Review of the research status of campus and urban boundary space. Urban Arch (27):42–44. https://doi.org/10.19892/j.cnki.csjz.2021.27.12

Zhang Y (2005) Comparative analysis of borderless campus and traditional campus teaching space. Cent China Constr S1:23–24

# Chapter 5
# Research on Supply Chain Management of Prefabricated Buildings Based on Bibliometrics

**Yuhang Zhang and Shengdong Cheng**

**Abstract** In order to understand the research content and research hotspots in the field of prefabricated building supply chain management, this paper uses the bibliometric method to sort out the domestic and foreign literature on prefabricated building supply chain management published from 2008 to 2021 through CiteSpace and VOSviewer software, and draws a scientific knowledge map. The results show that the number of published papers on the supply chain management of prefabricated buildings is generally on the rise. 'Journal of cleaner production' is a journal with great influence and reference value, and research in this field has formed a relatively obvious core author group, mainly concentrated in universities, forming two cooperation centers mainly in Hong Kong and mainland China. The research on the supply chain management of prefabricated buildings mainly focuses on the two aspects of cost and schedule. The future research focuses on the dynamic management of the supply chain and the coordination among stakeholders. The research frontier is to manage risks and coordinate the supply chain through the design platform.

**Keywords** Prefabricated buildings · Supply chain management · Bibliometrics · Citespace · VOSviewer

## 5.1 Introduction

Prefabricated building is an important way of building industrialization and the core means of transformation and upgrading of construction industry. The policy points out that it is necessary to vigorously develop prefabricated buildings, promote the formation of a complete industrial chain, improve the level of building industrialization, and strive to reach more than 30% of new prefabricated buildings by 2025 [1]. At this stage, under the policy-driven and market-led, prefabricated buildings

Y. Zhang · S. Cheng (✉)
State Key Laboratory Base of Eco-hydraulic Engineering in Arid Area, Xi' an University of Technology, Xi' an 710048, China
e-mail: chengsd@xaut.edu.cn



P. Mendonça et al. (eds.), *Proceedings of 2023 International Conference on Green Building*, Environmental Science and Engineering,
https://doi.org/10.1007/978-3-031-43478-5_5

have developed rapidly, and the supply chain has taken shape [2]. The prefabricated building supply chain is different from the traditional building supply chain. The prefabricated building supply chain has a short construction cycle, it is produced according to the order in the form of factory prefabrication and on-site installation; and it involves many participating enterprises, complex majors, information sharing and high coordination [3]. Prefabricated building supply chain management is an important way for prefabricated building to improve to plan, organize, coordinate and control the material flow, information flow and capital flow of prefabricated components It is combined with advanced technology in the process of ordering, production, transportation and assembly [4]. Prefabricated building supply chain management is an important way to improve production efficiency and competitiveness, to control costs and schedules, to save energy and reduce emissions [5]. However, problems such as improper way of storage and transportation of prefabricated components, and untimely information transmission have resulted in low supply chain efficiency and high cost. Supply chain management plays a key role in prefabricated buildings. If mismanagement occurs, the cost savings brought by economies of scale and the time saved by rapid installation will easily disappear [6]. At present, the supply chain management of prefabricated buildings is faced with information fragmentation, poor traceability, and lack of real-time information. At the same time, the temporary nature of projects will hinder the establishment of trust, making participants reluctant to share information. Therefore, the research of prefabricated building supply chain management is of great significance to achieve the goal of assembly building project and improve the efficiency of supply chain.

There are few researches on prefabricated building supply chain management in China, and there are certain limitations in the selection of analysis methods and objects. This paper uses the bibliometric method, selects CiteSpace and VOSviewer software to scientifically measure the literature related to the supply chain management of prefabricated buildings selected from the core collection database of China Academic Journal Full-text Database (CNKI) and Web of Science (WOS), analyzes the development status, scientifically and quantitatively summarizes and analyzes the research hotspots and frontiers, in order to provide reference for the research of supply chain management of prefabricated buildings in China and the promotion of prefabricated buildings.

## 5.2 Data Sources and Research Methods

### *5.2.1 Data Sources*

In order to cover the research status in the field of supply chain management of prefabricated buildings as comprehensively as possible, this paper used the Chinese Academic Journal Full-text Database (CNKI) and the Web of Science Core Collection (WOS) database as the retrieval platform, set the retrieval method as topic retrieval, considered the difference between Chinese and English expressions and determined the search term. Since journal papers can provide more comprehensive and high-quality information, this paper only analyzes published journal papers and excludes conference papers, book reviews, books, newspapers and letters which retrieval time is 2008–2021 to ensure the quality and reliability of the analysis. The literatures retrieved from CNKI and WOS were sorted out respectively, and the repetitive literatures and literatures inconsistent with the theme were eliminated. After screening, 62 literatures from CNKI and 102 literatures from WOS were obtained.

### 5.2.2 *Research Methods*

First of all, through the Statistical Analysis Toolkit for Informetrics (SATI) literature title information statistical analysis tool combined with the bibliometric analysis function of the WOS and CNKI databases, the annual publication volume, key journals, core authors, research institutions and countries of the literature in the field of prefabricated building supply chain management research are descriptively analyzed, and the field is sorted out and summarized as a whole. Secondly, through the bibliometric analysis software VOSviewer (1.6.17 version) and CiteSpace (5.8.R2 version), the research status, research hotspots and frontiers in the field of prefabricated building supply chain management are summarized and analyzed, and the results are visualized.

## 5.3 Knowledge Graph Analysis

### 5.3.1 *Time Distribution*

Scientific literature is the expression of academic research results. The number and change of published literature are important indicators to reflect the development level of a discipline [7]. This can be seen in Fig. 5.1. The number of papers published from 2008 to 2021 is on the rise as a whole, and the development and accumulation of knowledge in the field of supply chain management for prefabricated buildings is generally growing rapidly. According to the growth in the number of documents can be roughly divided into two stages.

(1) From 2008 to 2016, with the development of prefabricated buildings, the number of papers on supply chain management of emerging prefabricated buildings has grown from scratch. The number of papers at this stage is small, and the annual number of papers is below 5, indicating that the research on supply chain management of prefabricated buildings is in the early stage of development.

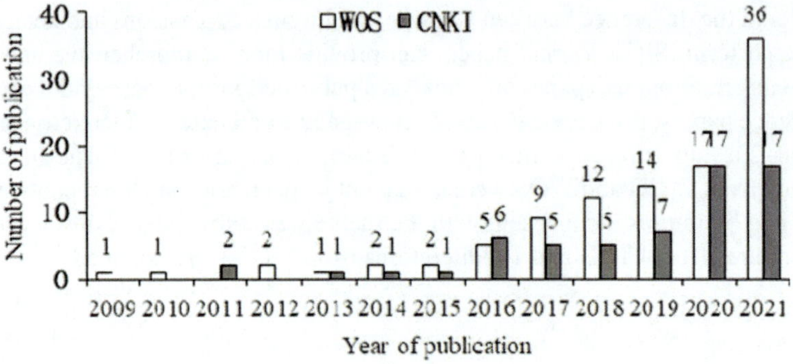


**Fig. 5.1.** Annual distribution of CNKI and WOS literatures

(2) From 2016 to 2021, the number of documents will increase year by year from 2016, and the growth rate will accelerate. The main reason is that in September 2016, the General Office of the State Council issued the 'Guiding Opinions of the General Office of the State Council on Vigorously Developing Prefabricated Buildings' [8]. Under the guidance of national policies, the promotion of prefabricated buildings has accelerated, and the problems in the field of prefabricated building supply chain have received more and more attention. The demand for prefabricated building supply chain management theory is growing. With the rise of smart construction sites [9] and intelligent supply chains in the future, the number of publications in the field of prefabricated building supply chains is expected to continue to grow in the future.

### 5.3.2 *Journal Distribution*

Key journals are an important window to explore the research progress and development trend in this field. They are also the focus of future research to understand the latest research trends in this field. They are crucial for researchers to find research channels and publish relevant research results [10]. There are 38 journals in the WOS literature in the field of prefabricated building supply chain management, with a total of 50 articles published, accounting for nearly half. There were 19 journals in CNKI literature, and 42 articles were published, accounting for 68.29%.

'Journal of cleaner production' has the largest number of publications, the highest impact factor (impact factor 9.44) in the past 5 years, and the highest H index (H index 200), indicating that the journal has great influence and reference value in this field. The impact factors of 'Journal of construction engineering and management' and 'Sustainability' were 4.513 and 3.473 respectively, and the H indexes were 114 and 85 respectively. 'Building Economy' is the journal with the largest number of articles, with 31 articles, accounting for half of the total number of documents. In

2021, the composite impact factor is 1.695, which has great influence and reference value in this field.

### *5.3.3 Scientific Collaboration Analysis*

The level of literature output reflects the academic ability and contribution of researchers to a certain extent. Through the analysis of scholar cooperation network, it can identify and present the cooperative relationship between the main research teams and researchers in a research field. The network diagram of research institutions can reflect the connection and contribution of different institutions in the field of academic research. The 102 papers in the WOS database were imported into the VOSviewer software, and statistics on the top 5 scholars and institutions by publication volume are shown in Table 5.1.

In the VOSviewer software, drawing scholar collaboration network diagram and institutional collaboration network diagram. As shown in Figs. 5.2 and 5.3, according to Price's law [11] core author statistics: $M = 0.749\sqrt{N_{\max}} = 0.749 \times \sqrt{9} = 2.247$, that is, the number of published papers of 3 or more as the core author. In this paper, a total of 20 core authors, published 81 articles, accounting for 79.41% of the total literature. The core authors of the top five are relatively concentrated, mainly forming eight cooperative groups dominated by Shen, Geoffrey Qiping, Xue, Fan and Zhai, Yue, etc. There is a certain degree of cooperation among the groups, and a relatively stable core author group has been formed. More than half of the scholars in the research group are from China, which has played a certain role in promoting the research level in the field of prefabricated building supply chain management in

**Table 5.1** Top 5 Statistics for WOS publications

| Scholars | Number of pub | Total citations | Institutions | Number of pub | Total citations |
|---|---|---|---|---|---|
| Shen, Geoffrey Qiping | 9 | 444 | Hong Kong Polytechnic University | 19 | 678 |
| Xue, Fan | 7 | 288 | University of Hong Kong | 18 | 409 |
| Li, Clyde Zhengdao | 5 | 339 | Shenzhen University | 10 | 360 |
| Zhai, Yue | 5 | 78 | Beijing Jiaotong University | 7 | 120 |
| Huang, George Q | 4 | 32 | Shanghai Jiao Tong University | 6 | 223 |

China. The publishing institutions are mainly concentrated in colleges and universities and are relatively concentrated. They are mainly Hong Kong Polytechnic University and Hong Kong University in Hong Kong. The total number of publications issued by the two universities accounts for 36.28%. Hong Kong has a small area and high construction cost, so it urgently needs the theory of prefabricated building supply chain management [12].

Through the CiteSpace software, the country cooperation network analysis, the number of papers in the leading position is China 57, accounting for 55.88%; 18 in the United States, accounting for 17.65%; Australia 14, accounting for 13.73%. As the largest node, China cooperates closely with other countries. China In recent years, due to the disappearance of the demographic dividend and high-quality development requirements [13], the construction industry must be transformed, mass production and modular technology to develop prefabricated buildings to solve the current housing problem.

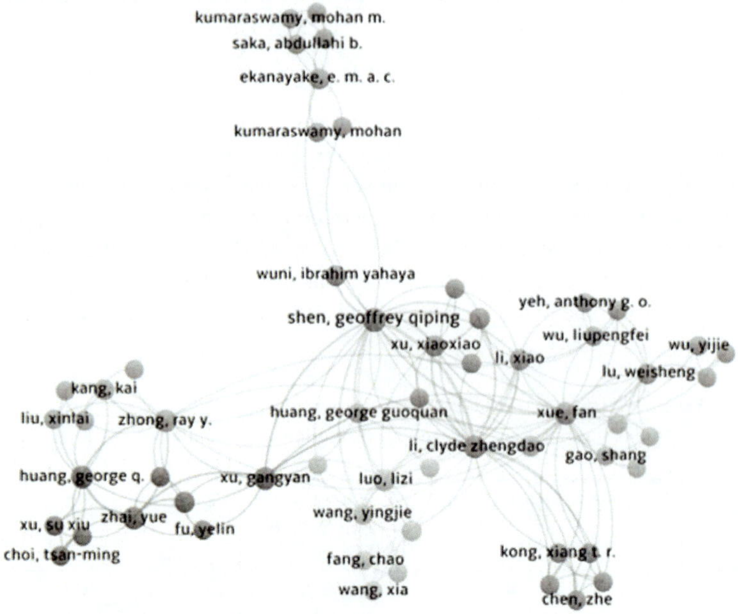


**Fig. 5.2** WOS scholar collaboration network

**Fig. 5.3** WOS research collaboration network

## 5.4 Research Hotspots and Frontier Analysis

### *5.4.1 High Frequency Keyword Analysis*

The key words of the literature are the high generalization and conciseness of the content of the literature, which can highly reflect the theme concepts and ideas of the article. SATI software was used to analyze the keywords of the literature. 183 keywords were obtained from CNKI literature and 353 keywords were obtained from WOS literature. In order to accurately understand the research hotspots in the field of prefabricated building supply chain management, remove search terms and general terms, compare the connotation similarity of keywords, and classify and merge keywords with consistent conceptual connotations. Finally, the top 10 high-frequency keywords are obtained in Table 5.2.

From Table 5.2, it can be found that the assembly building supply chain management focuses on cost overruns and schedule delays. In terms of schedule delay, RFID technology can track prefabricated components, master real-time information, dynamic planning, and improve supply chain performance. Uncertainty is an important source of schedule delay, and the probability of supplier uncertainty is small. For cost overruns, previous studies have focused on reducing costs by optimizing procurement decisions. However, decision parameters have not yet been developed. At present, the relevant research establishes a test platform through model assumptions, analyzes the impact of decision-making on costs, and proposes an optimization model to improve supply chain performance with a smaller total investment. At the

**Table 5.2.** Top 10 high frequency keywords

| WOS | | | CNKI | | |
|---|---|---|---|---|---|
| Number | Keywords | Frequency | Number | Keyword | Frequency |
| 1 | Performance | 24 | 1 | BIM | 4 |
| 2 | System | 19 | 2 | Influencing factor | 4 |
| 3 | Model | 16 | 3 | Whole industrial chain | 4 |
| 4 | Design | 14 | 4 | Green supply chain | 3 |
| 5 | Framework | 14 | 5 | Cost management | 3 |
| 6 | Optimization | 11 | 6 | Precast component | 3 |
| 7 | Barriers | 10 | 7 | Construction enterprise | 2 |
| 8 | Hong Kong | 10 | 8 | Lean construction | 2 |
| 9 | Simulation | 9 | 9 | RFID | 2 |
| 10 | China | 9 | 10 | Topsis | 2 |

same time, the designation of policies and plans at the government and industry levels has a significant impact on cost control.

### 5.4.2 *Cluster Analysis of Research Hotspots*

Clustering analysis is a common technique for statistical data analysis and knowledge discovery, which is used to identify semantic topics hidden in text data. Through statistical words and words in the same literature frequency of these words hierarchical clustering analysis to clarify the research status of prefabricated construction supply chain management. In VOSviewer software, keywords as nodes, draw WOS keyword co-occurrence network map, as shown in Fig. 5.4. The WOS keyword co-occurrence network diagram is divided into four categories. Among them, class 1 mainly studies the risk factors such as schedule, quality, stakeholders and the interaction between risk networks in the supply chain management of prefabricated buildings, and uses social network analysis methods and BIM and RFID technologies to effectively and efficiently track and control risks and improve project performance. Class 2 mainly studies the related problems of prefabricated components, and studies the dynamic scheduling of multiple production lines based on genetic algorithm to promote the on-time delivery of prefabricated components. Class 3 mainly discusses supply chain coordination. Class 4 mainly studies how to improve supply chain performance through the development of related technologies and platforms, and coordinate various stakeholders to achieve a win-win situation.

The keywords of Block-chain, platform, risks, circular economy, coordination and sustainability appear in 2021, indicating that the research frontier in the field

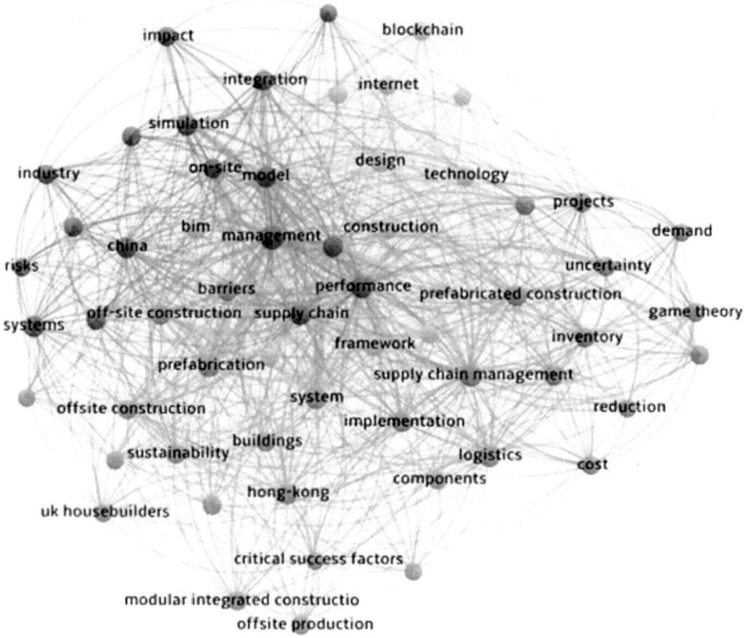


**Fig. 5.4** WOS keyword co-occurrence network

of prefabricated building supply chain management mainly involves risk management and coordination through blockchain technology and design management platform with circular economy as the main goal under the requirements of sustainable development.

## 5.5 Conclusion

This study uses the methods of mathematical statistics and bibliometrics to analyze the literature on the research of assembly building supply chain management from China Academic Journal Full-text Database (CNKI) and Web of Science Core Collection Database (WOS), and summarizes the research content, development trend and frontier hotspots of prefabricated building supply chain management with the help of knowledge map. The following conclusions are drawn:

(1) From 2008 to 2021, the annual publication volume of the research field of prefabricated building supply chain management is generally on the rise, which can be divided into two stages: early start-up development and rapid growth. With the construction of intelligent construction sites and intelligent supply chains, the number of publications in the field of prefabricated building supply chains will continue to grow in the future.

(2) The popular journals in the field of prefabricated building supply chain management mainly include 'Journal of cleaner production', 'Journal of construction engineering and management' and 'Automation in construction'. 'Journal of cleaner production' is a journal with great influence and reference value in this field. Domestic popular journals are mainly 'Construction Economy' and 'Journal of Civil Engineering and Management'.
(3) At present, the relevant research on the supply chain management of prefabricated buildings has formed a relatively obvious core author group, but lack of domestic and foreign cooperation, and it is necessary to deepen exchanges and cooperation. The research institutions are mainly concentrated in universities, forming two major cooperation centers in Hong Kong and mainland China. At the national level, China, the United States and Australia have more research results and higher influence.
(4) The research of prefabricated building supply chain management mainly focuses on cost and schedule. The future research will focus on the dynamic management of supply chain and coordination among stakeholders. The research frontier is to manage risk and coordinate supply chain through design platform.

**Acknowledgements** This research was supported by the National Natural Science Foundations of China (No. 42107183).

## References

Ministry of Housing and Urban-Rural Development (2022) This year will vigorously develop prefabricated buildings. Concrete (2):79
Zengke Y, Ruiguo F, Wei H, Ying S (2022) Research on cooperation strategy of core enterprises in prefabricated building industry chain under government intervention. China Manag Sci 1–11
Tao Z, Yaping Z, Yuchen G (2022) Multi-dimensional interpretation of prefabricated building industry chain and evaluation of influencing factors of supply chain autonomy and controllability. J Archit Sci Eng 1–13
Zhang X, Skitmore M, Peng Y (2014) Exploring the challenges to industrialized residential building in China. Habitat Int 41
Liu Y, Dong J, Shen L (2020) A conceptual development framework for prefabricated construction supply chain management: an integrated overview. Sustainability 12(5)
Junyoung J, Seungjun A, Hyun CS, Kyeongwoon C, Choongwan K, Wan KT (2021) Toward productivity in future construction: mapping knowledge and finding insights for achieving successful offsite construction projects. J Comput Des Eng 8(1)
Lingzhi Li, Dong Chen, Ling Shen (2021) Knowledge map visualization analysis of infrastructure resilience assessment research. J Fudan Univ (Nat Sci Ed) 60(1):14–26
Wenjun M (2021) Research on the whole process quality risk assessment of prefabricated buildings. Southeast University
Jiancheng H, Kun X, Zhanbo D (2021) Research and implementation of intelligent site management platform system architecture. Build Econ 42(11):25–30
Prince A-A, Thomas NS, Md. Uzzal H (2021) A review of the circularity gap in the construction industry through scientometric analysis. J Clean Prod 298

Hui M, Meihong D, Weiwen W, Bingqian H (2020) Research progress and trend of construction industrialization at home and abroad—visual analysis based on Cite space. J Civil Eng Manag 37(1):43–49 + 56

Luo L, Jin X, Shen GQ, Wang Y, Liang X, Li X, Li CZ (2020) Supply chain management for prefabricated building projects in Hong Kong. J Manag Eng 36(2)

Du Q, Pang Q, Bao T, Guo X, Deng Y (2021) Critical factors influencing carbon emissions of prefabricated building supply chains in China. J Clean Prod 280(Pt 2)

# Chapter 6
# Embodied Carbon Footprint Analysis of Signage Industry: Insights from Two Case Studies

**Prudvireddy Paresi, Fatemeh Javidan, and Paul Sparks**

**Abstract** Embodied carbon has recently become a hot topic among environmentalists and designers, especially after the Paris Agreement on climate change. Embodied carbon refers to the carbon emissions associated with the manufacturing and transportation of building materials and the process of construction. The "Global Status Report for Buildings and Construction" report estimated that the building and construction sector alone contributed nearly 37–39% of global carbon emissions in 2017–2020. To tackle embodied carbon, the World Green Building Council (WorldGBC) has set a bold vision to reduce it by at least 40% by 2030 and achieve net-zero operating carbon in all new buildings. The signage industry plays a significant role in the building industry, as signages are a key component of buildings. Signages serve multiple purposes, such as providing information, enhancing brand identity, and promoting safety. Therefore, it is essential to understand the embodied carbon emissions associated with signage materials used to minimise the overall carbon emissions of construction projects. The present paper aims to study the embodied carbon footprint of the signage industry with the help of two case studies. The embodied carbon factors required while estimating the overall footprint of the signages are taken from Environmental Performance in Construction (EPiC) database. The study identifies the aluminum as the major contributor of the embodied emissions in the signage projects. This study provides insight into the other sources of embodied carbon and makes more informed decisions while selecting signage materials used in designs to create sustainable and economic projects. This information helps to increase sustainability and reduce the carbon footprint of signage projects in the early decision-making stages.

P. Paresi (✉)
Global Professional School, Federation University, Ballarat, Australia
e-mail: p.paresi@federation.edu.au

F. Javidan
Institute of Innovation, Science and Sustainability, Federation University, Ballarat, Australia

P. Sparks
Sustainability, Diadem Pty Ltd, Melbourne, Australia



P. Mendonça et al. (eds.), *Proceedings of 2023 International Conference on Green Building*, Environmental Science and Engineering,
https://doi.org/10.1007/978-3-031-43478-5_6

**Keywords** Signage industry · Embodied carbon · Carbon-intensive materials · Carbon emissions

## 6.1 Introduction

The Paris Agreement sets targets to limit the rise in global average temperatures well below 2 °C above pre-industrial levels (UNFCC 2015). Buildings play a major role in the production of greenhouse gases (GHGs), which ultimately contribute to higher temperatures and a more severe impact of climate change. Generally, two kinds of carbon emissions are associated with building materials: operational carbon and embodied carbon (Ibn-Mohammed et al. 2013). Operational carbons are associated with energy usage for building operations such as heating, cooling, lighting, maintenance, etc. Embodied carbons are associated with the extraction, manufacturing, transportation, installation and repair of materials or associated products (Ibn-Mohammed et al. 2013). In 2020, buildings alone are responsible for 39% of the global energy-related carbon emissions, of which 28% are operational, and 11% are embodied carbon (U.N.E. Programme 2021). Although operational carbon has played a major role in total world carbon emissions over the last few decades, building's long lifecycles, recent advancements in the renewable energy sector, and insulation methodologies provide promising opportunities to control operational carbon. On the other hand, if no action is taken to reduce embodied carbon, its shares are projected to reach 85% by 2050 (GBCA and Thinkstep-anz 2021).

The signage industry involves designing, manufacturing, and installing various types of signs, including wayfinding signs, interior and outdoor signs, vehicle graphics, and trade show displays. These signages serve a variety of functions, including advertising, brand promotion, direction-giving, and information display. Printed signage and digital signage are the two major categories in the signage industry. Posters, banners, yard signs, decals, and stickers are some examples of printed signage. On the other hand, digital signage, such as LED/LCD displays, video walls, digital posters etc., are also highly adopted in the building industry due to their ease in customising signs with unique designs, images, and messages (Sepasgozar and Davis 2019). The development of new technologies and materials has allowed for manufacturing of high-quality, long-lasting signs that are more resilient, vivid, and affordable than ever before, which has led to a considerable increase in the signage sector in recent years. The majority of the building materials are widely used in the signage industry. The materials chosen to fabricate signs are crucial in determining their overall aesthetic appeal and durability. For example, aluminium is a popular choice due to its lightweight, corrosion-resistant properties and ability to form intricate shapes easily. Meanwhile, steel is frequently used to provide the necessary strength and durability, especially for supporting structures.

However, signage materials such as steel, aluminium, concrete, acrylic and vinyl are highly energy-containing materials. They consume a lot of energy during the extraction and fabrication stages. Selecting materials with high embodied energy results in higher embodied carbon emissions (Zeng and Chini 2017). Consequently, selecting these signage materials plays a significant role in reaching sustainability goals. Since most of these embodied energies and embodied carbon are created before the signage is built, choosing materials with low embodied energy and embodied carbon during the design stage of the signage makes a big difference to the sustainability of the signage projects (Bribián et al. 2011).

## 6.2 Methodology

The process of selecting the two cases for this study involved careful consideration of various factors. Firstly, the case studies should represent typical scenarios within the signage industry, allowing for a comprehensive analysis of embodied carbon emissions. Secondly, the chosen case studies should cover a range of signage types and materials commonly used in practice. This ensured the findings would have broader applicability and relevance to the industry as a whole. To estimate the overall embodied carbon emissions in the project, there is a need for a precise and reliable carbon factor database. Multiple research groups and institutions have been working to develop such datasets for construction materials, particularly for the cradle-to-gate stage. To enhance accuracy, this study employs the Environmental Performance in Construction (EPiC) database (Crawford et al. 2019), based on material data obtained locally in Australian markets, to derive embodied carbon emission factors. Table 6.1 displays the embodied carbon factors used in this present study.

In the present study, the total embodied carbon emissions are estimated using the following mathematical formula:

$$\boldsymbol{EC_{SM}} = \sum_{i=1}^{n} \boldsymbol{Q_n . e_{nsm}} \tag{6.1}$$

where, $\boldsymbol{EC_{SM}}$ is the embodied carbon emission (in $kgCO_2e$) of signage material (sm) used, $\boldsymbol{Q_n}$ is the amount of n-th signage material used (in kg) and $\boldsymbol{e_{nsm}}$ is the emission factor for the n-th signage material ($kgCO_2e/kg$).

**Table 6.1** Embodied carbon factors of the signage materials used in this work

| Material | Embodied carbon factor |
|---|---|
| Mild steel | 3.7 (Crawford et al. 2019) |
| Aluminum | 12.8 (Crawford et al. 2019) |
| Vinyl | 390 (Furney 2019) |
| Acrylic PMMA | 15.4 (Crawford et al. 2019) |
| Concrete 25 MPa | 0.8 (Crawford et al. 2019) |

### 6.2.1 Case Study 1: Retail Banking Signage

Case study 1 estimates the embodied carbon for a free-standing pylon double-sided illuminated (FSP2s) sign in front of a Commonwealth Bank of Australia branch under a test-and-learn scenario. This signage is attached to a 100 × 50 × 3 mm galvanised mild steel rectangular hollow section (RHS) bar, which stands upright and fixed between the ground and ceiling at the branch's entrance. The height of the RHS bar can vary from branch to branch, depending on the roof height. A square aluminium perimeter frame is welded to the 50 × 3 mm SHS aluminium bar, and in addition to this perimeter frame, a 32 × 3 mm aluminium equal angle (EA) frame is welded to the SHS bar. A square-shaped 4.5 mm acrylic moulding with a 16μ digital printed beacon (vinyl) is screwed between the aluminium perimeter frame and the EA frame all around the circumference of the frame. LED strips are fixed in the middle of the frame for illumination purposes.

### 6.2.2 Case Study 2: Convention Center Signage

Case study 2 estimates the embodied carbon emissions of monopole free-standing double-sided non-illuminated (MFD) signs in front of a convention center to provide directions and information to pedestrians or vehicular traffic. The production of MFD signs typically involves a combination of materials like aluminium, steel, concrete, vinyl, and acrylic. In the present study, the MFD sign comprises two finger blade boards made from a 6 mm thick aluminium plate with customised dimensions. The site directions are inscribed on top of these boards using vinyl material. The sign pole, made of galvanised steel and has dimensions of 100 × 4 mm square hollow section (SHS), is attached to a concrete base with a volume of 0.36 $m^3$ via a base plate.

## 6.3 Discussions

The materials used in the signs for case study 1 and case study 2 are summarised in Tables 6.2 and 6.3, respectively, showing their corresponding mass (in kgs) and embodied carbon emissions ($kgCO_2e$). Based on Table 6.2, in case study 1, steel contributes the highest mass at 22.8 kg, followed by aluminium and acrylic PMMA at 15.8 kg and 11.1 kg, respectively. At the same time, aluminium contributes to 202.2$kgCO_2e$, followed by acrylic PMMA and mild steel at 170.5 and 93.9 $kgCO_2e$. Similarly in case study 2 (from Table 6.3), it can be observed that concrete contributed to 867.2 kg of mass followed by steel and aluminum of 66 and 13.8 kg. In term of embodied carbon, steel has the highest contribution (266.9 $kgCO_2e$), followed by aluminium (176.8 $kgCO_2e$) and concrete (130 $kgCO_2e$). These findings suggest that steel, aluminium, concrete, and acrylic PMMA are the predominant materials used in signage projects. Therefore they significantly impact the overall embodied carbon emissions of the signage projects.

Figure 6.1 displays the % mass and % embodied carbon contribution of the signage materials used in case study 1. It can be observed that steel, which has the highest contribution to the overall mass, accounts for 19.6% of the embodied carbon. Despite its lower mass, aluminium (42.3%) and acrylic PMMA (35.6%) contribute to higher embodied carbon than steel. Figure 6.2 displays the percentage mass and embodied carbon contribution of different materials used in case study 2. It can be seen that concrete, regardless of its highest mass contribution (91.6%) compared to steel (7%) and aluminium (1.5%), has contributed to lower embodied carbon due to its lower embodied carbon factor.

In both case studies, aluminium, despite lower mass shares, contributes relatively high embodied carbon due to its high embodied carbon factor. It can be attributed to its energy-intensive production process, which requires significant energy to separate aluminium from the oxygen in the alumina (Brough and Jouhara 2020). Consequently, it emits substantial greenhouse gases, resulting in a higher embodied carbon value. It can also be observed that vinyl's contribution to embodied carbon is relatively low (2.4%), despite having a very high embodied carbon factor than all other signage materials due to its very low mass contribution (0.06%). It emphasises the importance of considering material mass and embodied carbon during material selection for signage projects.

**Table 6.2** List of materials and their mass and embodied carbon contributions in case study 1

| Part name | Material | Mass (kg) | Embodied carbon (kg$CO_2$e) |
|---|---|---|---|
| MS RHS spigot | Steel | 16.5 | 67.7 |
| MS base plate | Steel | 6.3 | 26.2 |
| Aluminum_SHS | Aluminum | 5.2 | 67.1 |
| Aluminum EA | Aluminum | 7.4 | 94.2 |
| Aluminum_Frame | Aluminum | 3.2 | 41.0 |
| Acrylic moulding | Acrylic PMMA | 11.1 | 170.5 |
| Digital beacon | Vinyl | 0.03 | 11.6 |

**Table 6.3** List of materials and their mass and embodied carbon contributions in case study 2

| Part type | Material | Mass (kg) | Embodied carbon (kg$CO_2$e) |
|---|---|---|---|
| Vertical support | Steel | 56.8 | 232.9 |
| Base plate and screws | Steel | 9.2 | 34 |
| Finger blade 1 | Aluminum | 5.8 | 73.9 |
| Finger blade 2 | Aluminum | 8 | 102.9 |
| Footage | Concrete | 867.2 | 130 |
| Graphics (labels) | Vinyl | 0.004 | 1.8 |

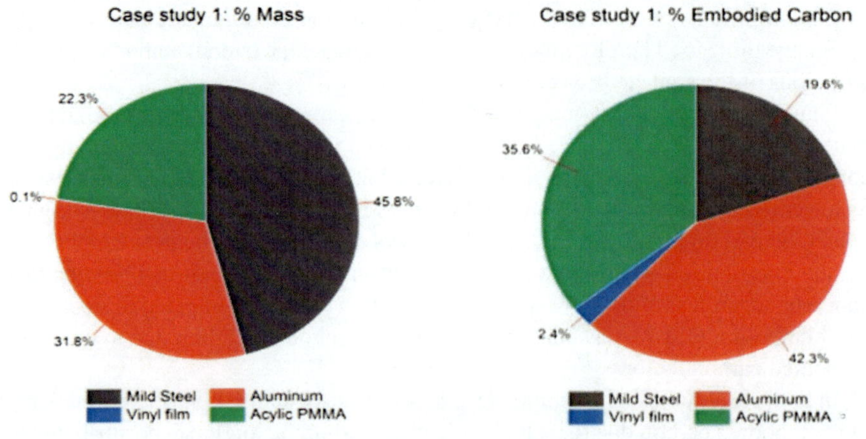


**Fig. 6.1** % Mass and % embodied carbon contributions of signage materials used in case study 1

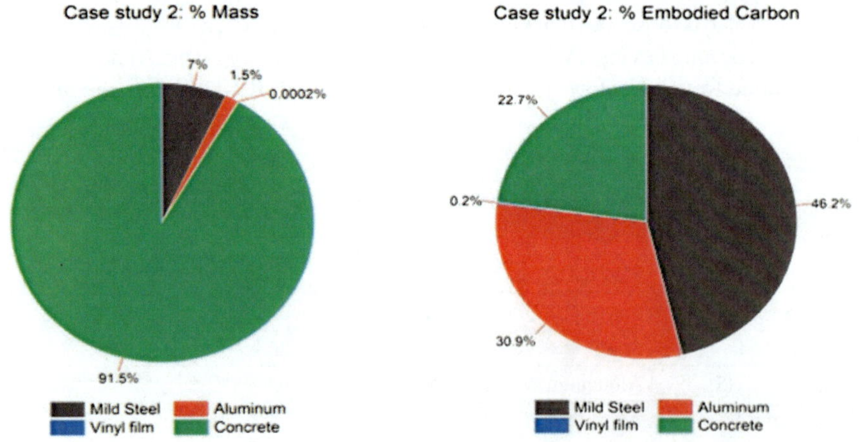


**Fig. 6.2** % Mass and % embodied carbon contributions of signage materials used in case study 2

Additionally, the embodied carbon of concrete in case study 2 is lower than steel, despite its higher mass contribution. This can be attributed to the lower embodied carbon factor of concrete, which implies that its production emits less carbon than steel production. This demonstrates the importance of selecting materials with low embodied carbon factors to minimise the overall environmental impact without compromising the functionality and aesthetics of signage projects.

Therefore, to mitigate the embodied carbon emissions in signage projects, it is important to employ low embodied carbon materials or to minimise the mass of higher embodied carbon materials. Opting for green materials such as timber and bamboo, which have a lower embodied carbon factor than traditional materials like steel and aluminium, can also be a better alternative. Sustainably sourced timber accounts for negative carbon emissions (Hammond and Jones 2019). Similarly, replacing conventional materials with recycled materials can also effectively lower embodied carbon emissions. However, the choice of materials should not compromise the structural integrity and safety of the signs. It is important to note that the embodied carbon values reported in most of the databases, including EPiC, represent only the carbon emissions from the cradle-gate phase of each material and do not account for emissions from transportation, installation, or disposal. Therefore, the embodied carbon values should be interpreted with caution when comparing the environmental impact of different materials.

## 6.4 Conclusions

In conclusion, the present study highlights the importance of estimating embodied carbon in the signage industry with the help of two case studies. The study used the EPiC database, developed from local material sources, to derive embodied carbon emission factors for the signage materials analysed. Results indicate that aluminium, steel, acrylic PMMA, and concrete are the most common materials used in the signage industry. Being highly carbon intensive, these materials are responsible for a significant portion of the carbon emissions associated with signage projects. The study also reveals that despite having a lower mass contribution due to its lower density, aluminium's high embodied carbon factor leads to a significant contribution to the overall embodied carbon emissions. Designers and industry professionals must consider more sustainable alternatives for these materials, especially aluminium. The study also suggests sustainably sourced timber or recycled materials as viable and sustainable alternatives to aluminium. By opting for these alternatives, we can reduce the environmental impact of the signage industry and create more sustainable projects. The quantification of the embodied carbon emissions and economic costs associated with these low carbon materials requires further investigation, which is planned as part of future work.

**Acknowledgements** The authors acknowledge the Commonwealth's support presented by the Department of Industry, Science, Energy and Resources through the Innovation Connections grant ICG001786 and Diadem DDM PTY LTD's support as the project partner. We would further like to thank "Commonwealth Bank of Australia" for their support during this research study.

## References

Bribián IZ, Capilla AV, Usón AA (2011) Life cycle assessment of building materials: comparative analysis of energy and environmental impacts and evaluation of the eco-efficiency improvement potential. Build Environ 46(5):1133–1140

Brough D, Jouhara H (2020) The aluminium industry: a review on state-of-the-art technologies, environmental impacts and possibilities for waste heat recovery. Int J Thermofluids 1:100007

Crawford RH, Stephan A, Prideaux F (2019) Environmental performance in construction (EPiC) database

Furney J (2019) 3M LX480 environmental product declaration

GBCA and Thinkstep-anz (2021) Embodied carbon and embodied energy in Australia's buildings. Green Building Council of Australia and thinkstep-anz, Sydney

Hammond G, Jones C (2019) Inventory of carbon & energy: ICE v3, vol 5. Sustainable Energy Research Team, Department of Mechanical Engineering.

Ibn-Mohammed T et al (2013) Operational vs. embodied emissions in buildings—a review of current trends. Energy Build 66:232–245

Sepasgozar SM, Davis S (2019) Digital construction technology and job-site equipment demonstration: modelling relationship strategies for technology adoption. Buildings 9(7):158

U.N.E. Programme (2021) Global status report for buildings and construction. Global Alliance for Building and Construction

UNFCC (2015) Key aspects of the Paris Agreement. https://unfccc.int/resource/docs/2015/cop21/eng/10a01.pdf

Zeng R, Chini A (2017) A review of research on embodied energy of buildings using bibliometric analysis. Energy Build 155:172–184

# Chapter 7
# Energy Efficiency Renovation Packages for European Supermarkets: The Experience of SUPER-HEERO Project

**Giorgio Bonvicini, Sara Abd Alla, Nora Ganzinelli, Cristina Barbero, and Thomas Messervey**

**Abstract** In 2020, buildings accounted for 40% of EU final energy consumption and 36% of GHG emissions, a significant share of which is due to supermarkets, which are among the most energy intensive buildings. Supermarkets are also interesting for their potential to serve as energy transition change agents due to their role in the social fabric of local communities (e.g.: influencing citizens to make sustainable actions and choices). To reduce energy consumption in supermarkets, the EU co-funded SUPER-HEERO project is developing a replicable financial scheme for energy efficiency investments in small/medium supermarkets, based on stakeholder engagement. In the Project, based on an analysis of typical supermarkets' energy uses, a catalogue of 42 energy efficiency interventions was created, covering overall energy management, improvement of energy supply and main energy uses, i.e. HVAC, lighting, products refrigeration, etc. One ambition is to demystify energy renovation packages and make supermarket stakeholders more likely to take action. In this paper, 18 energy efficiency renovation packages are created for supermarkets based on: baseline energy conditions, geographical location, renovation depth and associated investment needs. All KPIs for renovation packages are presented per unit of floor area to allow, with a certain degree of accuracy, extrapolation for supermarkets of different sizes. The work done demonstrates how a wide range of opportunities for energy efficiency-oriented renovation of supermarkets exist, characterized by a limited investment, a reduction of energy uses and GHG emissions of 20–30% and a payback below 5–6 years. The best cost–benefit ratio is found for "basic" renovation packages applied to old supermarkets, since they focus on "low-hanging fruits"; actions on supermarkets in better conditions might be less profitable but interesting and/or required to meet corporate strategies, especially engaging stakeholders like in SUPER-HEERO innovative financing scheme.

G. Bonvicini (✉) · S. Abd Alla · N. Ganzinelli
RINA Consulting S.p.A., Via A. Cecchi 6, Genova, Italy
e-mail: giorgio.bonvicini@rina.org

C. Barbero · T. Messervey
R2M Solution, Via Fratelli Cuzio 42, Pavia, Italy



P. Mendonça et al. (eds.), *Proceedings of 2023 International Conference on Green Building*, Environmental Science and Engineering,
https://doi.org/10.1007/978-3-031-43478-5_7

**Keywords** Supermarkets · Energy efficiency · Innovative financing

## 7.1 Introduction

According to the European Commission, buildings in 2020 were responsible for 40% of the EU energy consumption and 36% of greenhouse gas (GHG) emissions (European Commission 2021a). In December 2021, the European Commission adopted a legislative proposal to revise the Energy Performance of Buildings Directive (EPBD), as part of the "Fit for 55" package, and a new European Climate Law increasing the 2030 and the 2050 energy efficiency targets (European Commission 2021b).

Due to refrigeration, lighting and climatization demands, supermarkets are among the most energy intensive buildings (Acha et al. 2016). Although the feasibility of energy efficiency investments in supermarkets and other retail businesses has been widely demonstrated from technical, economic and environmental point of views, there are still barriers to the activation of energy efficiency investments by private investors, being them institutional, industry or commercial companies. Those barriers arise from the lack of confidence, limited knowledge of available technologies and little awareness of technical and financial risks. For this reason, there is a need to set up innovative financing schemes to make energy efficiency investments more attractive for private investors at different levels.

The SUPER-HEERO project, co-funded by the European Commission under the Horizon 2020 programme, aims at addressing the barriers to the implementation of energy efficiency investments in supermarkets, by developing a set of innovative and replicable financial schemes leveraging on stakeholders' engagement. The SUPER-HEERO project approach relies on three main instruments: engineered Energy Performance Contracts (EPC), "product-as-a-service" model for technology providers and community-based crowdfunding/cooperative initiatives. These innovative financing schemes may be combined into a hybrid composite solution (e.g.: crowdfunding to finance an EPC also employing as a service for part of a holistic intervention).

## 7.2 Energy Consumption in Supermarkets

Several studies are available in literature on energy consumption of supermarkets. Specifically, depending on the specific location, supermarkets have been identified to have an average final energy consumption in the range between 320 and 800 kWh/m$^2$y. Data taken from different sources (Lindberg 2018; Kolokotroni et al. 2019; SME Energy CheckUp Project 2012; SUPER Smart Project 2016; Sluis 2017) show that in Spain the average value was of 327 kWh/m$^2$y, in Sweden of 420 kWh/m$^2$y, in France of 570 kWh/m$^2$y, in Italy of 598 kWh/m$^2$y, in Poland of 800 kWh/m$^2$y, in

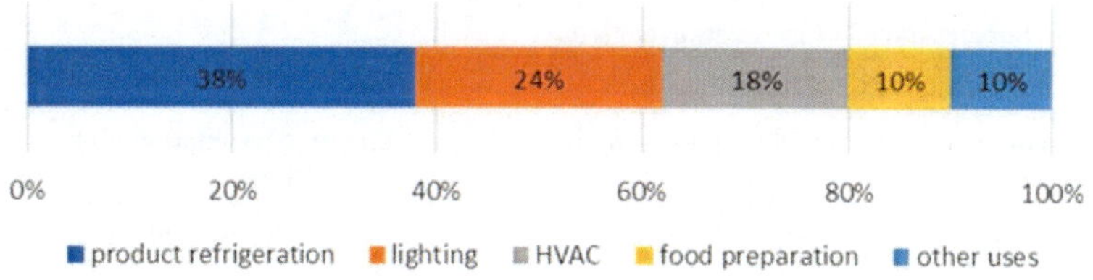


**Fig. 7.1** Breakdown of final energy uses in supermarkets, average values

the United Kingdom of 400–740 kWh/m$^2$y, in Norway of 500 kWh/m$^2$y and in the USA of 600 kWh/m$^2$y.

These final energy needs are almost always covered by electricity, except for the cases where boilers fed with natural gas or other fuels are used for heating and sanitary hot water production.

Considering the main energy flows in supermarkets that are well schematized in Lindberg (2018), the net energy inputs related to externally purchased or self-produced energy are used to cover energy demand related to space heating, cooling and products refrigeration, lighting, ventilation, sanitary hot water production and other appliances, including food preparation and office equipment.

From the same literature sources (Lindberg 2018; Kolokotroni et al. 2019; SME Energy CheckUp Project 2012; SUPER Smart Project 2016; Sluis 2017) an average breakdown of final energy uses among different areas/devices in the supermarket can be drawn and is presented in Fig. 7.1; it can be noticed that the largest share of energy use is due to product refrigeration (38%), followed by lighting (24%), HVAC (18%) and food preparation (10%); other energy uses cumulatively account for 10%.

## 7.3 Energy Efficiency Renovation Packages

The energy consumption of supermarkets strongly varies with building features, location and climate conditions as well as with the efficiency of the installed devices and the operational practices. Taking this into consideration, supermarkets have been clustered in this study based on baseline conditions (“old”, “average”, “new”), geographical location (“Northern” and “Southern” Europe) and budget availability and consequent renovation depth (“deep”, “partial” and “basic”) (SUPER-HEERO Project 2021a). The supermarkets’ clusterization criteria adopted in the SUPER-HEERO project are presented in Table 1 and will be the basis for the development of the proposed energy efficiency renovation packages.

Based on the supermarkets’ clusterization presented in Table 7.1, the 42 energy efficiency actions identified in SUPER-HEERO project have been analyzed and grouped; the most common interventions considered in the creation of the renovation packages are:

**Table 7.1** Supermarkets' clusterization criteria

| Clusterization criterion | Cluster | Indicator |
|---|---|---|
| Baseline conditions | Old with low-efficiency devices | Energy consumption higher than 700 kWh/m$^2$y |
| Baseline conditions | Average conditions | Energy consumption between 400 and 700 kWh/m$^2$y |
| Baseline conditions | Recent supermarket, updated technologies | Energy consumption lower than 400 kWh/m$^2$y |
| Climate conditions | Northern Europe | Latitude higher than 47° |
| Climate conditions | Southern Europe | Latitude lower than 47° |
| Budget availability | Deep renovation | CAPEX higher than 500 €/m$^2$ |
| Budget availability | Partial renovation | CAPEX between 200 and 500 €/m$^2$ |
| Budget availability | Basic renovation | CAPEX lower than 200 €/m$^2$ |

- LED lighting, applicable to all supermarkets using fluorescent lamps or other inefficient technologies, i.e. under the assumptions of this study, the "old" and "average" ones;
- installation of doors to reduce losses from refrigerated cabinets, which is recommended to "old" supermarkets where they typically are not present;
- fine-tuning of HVAC systems through the improved control of ventilation, hot/chilled water circulation and temperature/humidity control through installation of Variable Frequency Drives on electric motors, recommended to "old" and "average" supermarkets;
- conversion of space heating and cooling to high-efficiency heat pumps fed with electricity from renewable sources, recommended to "old" and "average" supermarkets in Southern Europe willing to undergo a "deep" or "partial" renovation;
- refurbishment of refrigeration systems, applicable to "old" and "average" supermarkets implementing a "deep" or "partial" renovation;
- thermal insulation of building envelope, applicable only in case of "deep" renovations to "old" and "average" supermarkets;
- cogeneration, applicable in Northern Europe for "old" and "medium" supermarkets;
- use of solar thermal modules for sanitary hot water production, applicable in Southern Europe to "new" or "average" supermarkets willing to implement a "deep" renovation;
- heat recovery from refrigeration plants to cover space heating demand or sell it to the local district heating network, applicable in Northern Europe to "new" supermarkets;
- smart control of electric loads (lighting, heating, ventilation and air conditioning—HVAC, also based on Artificial Intelligence solutions), recommended to "new" supermarkets with most of the basic energy efficiency actions already implemented;

- the installation of building-integrated renewable electricity production plants (mainly photovoltaic plants) is also an option, but subject to local pre-requisites (roof availability, orientation/slope, shadows, etc.), therefore this has not been included in renovation packages.

Based on the considerations above, 18 renovation packages are built, considering the three levels of renovation depth for the three level of baseline conditions and two geographical areas in Europe. The main features of the proposed renovation packages, including achievable energy savings, budget required and investment payback period, are presented in Table 7.2. The values presented in the Table are based on the experience of project partners in energy audits and feasibility studies for energy efficiency interventions, and on consultations with technology providers (SUPER-HEERO Project 2021a, b). The payback period is calculated by dividing the investment required by the annual economic savings achievable, given by the multiplication of the energy savings and the energy price (for electricity, natural gas, etc.).

The values presented in Table 7.2 may be affected by an uncertainty of approximately ±50%, due to several reasons including the investment needs not tailored on a specific site nor Country. Moreover, economic savings are calculated considering 150 €/MWh for electricity price and 30 €/MWh for natural gas price (i.e. those typical of the EU market before the 2022 energy crisis), which could not be representative of the specific market conditions at the renovation package implementation period.

## 7.4 Conclusions

The present paper demonstrates that opportunities for the improvement of energy efficiency of supermarkets exist and are characterized by a relatively limited investment and a significant reduction of energy uses and GHG emissions and consequently by a short payback time. These actions are grouped into 18 energy efficiency renovation packages, applicable to the typical supermarkets based on baseline conditions, geographical location and budget availability and consequent renovation depth.

When these renovation packages are applied to an average 400 $m^2$ supermarket, the results obtained show that investing less than 200,000 € a reduction of 20–30% of baseline energy uses can be obtained, therefore achieving investment payback in a 5–6 year period or lower, thanks to incentives.

Specifically, the lowest payback period is obtained for a "basic" renovation of an old supermarket, focusing on "low-hanging fruits"; actions on supermarkets in better conditions might be less profitable but still are interesting and/or required to achieve targets set at corporate level with reference to Environmental, Social and Governance (ESG) topics.

**Table 7.2** Main features of supermarkets' renovation packages

| # | Package description | Energy savings kWh/m$^2$y | Budget €/m$^2$ | Paybac years |
|---|---|---|---|---|
| 1a | Deep renovation, old supermarket, Northern Europe | 290.8 | 521.3 | 4.8 |
| 1b | Deep renovation, old supermarket, Southern Europe | 374.2 | 521.3 | 6.9 |
| 2a | Partial renovation, old supermarket, Northern Europe | 190.8 | 421.3 | 4.5 |
| 2b | Partial renovation, old supermarket, Southern Europe | 307.5 | 421.3 | 6.4 |
| 3a | Basic renovation, old supermarket, Northern Europe | 107.5 | 46.3 | 2.9 |
| 3b | Basic renovation, old supermarket, Southern Europe | 107.5 | 46.3 | 2.9 |
| 4a | Deep renovation, average supermarket, Northern Europe | 249.2 | 506.3 | 4.9 |
| 4b | Deep renovation, average supermarket, Southern Europe | 279.2 | 493.8v | 8.0 |
| 5a | Partial renovation, average supermarket, Northern Europe | 149.2 | 406.3 | 4.6 |
| 5b | Partial renovation, average supermarket, Southern Europe | 245.8 | 418.8 | 7.4 |
| 6a | Basic renovation, average supermarket, Northern Europe | 82.5 | 281.3 | 3.6 |
| 6b | Basic renovation, average supermarket, Southern Europe | 82.5 | 131.3 | 4.1 |
| 7a | Deep renovation, new supermarket, Northern Europe | 112.5 | 78.8 | 4.7 |
| 7b | Deep renovation, new supermarket, Southern Europe | 59.2 | 41.3 | 4.6 |
| 8a | Partial renovation, new supermarket, Northern Europe | 112.5 | 78.8 | 4.7 |
| 8b | Partial renovation, new supermarket, Southern Europe | 59.2 | 41.3 | 4.6 |
| 9a | Basic renovation, new supermarket, Northern Europe | 45.8 | 28.8 | 6.9 |
| 9b | Basic renovation, new supermarket, Southern Europe | 45.8 | 28.8 | 6.9 |

Although characterized by very good technical performances and an acceptable financial return, energy efficiency renovation packages could sometimes be difficult to get finance through conventional pathways, especially for small/medium supermarkets. For this reason, the innovative financing measures developed in the

SUPER-HEERO project, i.e. EPC contracts, product-as-a-service and crowdfunding, are particularly of interest to support the implementation of renovation packages.

Such packages can serve the dual purpose of assisting fleet managers to plan actions across sets of buildings and to make more accessible energy efficiency strategies for individual shop managers. Moreover, the bottom-up, "social and shared" innovation proposed, supports the implementation of energy efficiency actions since the community wants it, thus making payback a less important barrier.

**Acknowledgements** The SUPER-HEERO project has received funding from the European Union's Horizon 2020 research and innovation programme under Grant Agreement N° 894404. The opinion stated in this paper reflects the opinion of the authors and not the opinion of the European Commission.

## References

Acha S, Du Y, Shah N (2016) Enhancing energy efficiency in supermarket refrigeration systems through a robust energy performance indicator. Int J Refrig 64:40–50
European Commission (2021a) Making our homes and buildings fit for a greener future
European Commission (2021b) Regulation (EU) 2021b/1119 of the European parliament and of the council of 30 June 2021 establishing the framework for achieving climate neutrality and amending regulations (EC) no 401/2009 and (EU) 2018/1999
Kolokotroni M et al (2019) Supermarkets energy use in the UK. Energy Procedia 161:325–332
Lindberg U (2018) Exploring barriers to energy efficiency in supermarkets. PhD thesis, University of Boras, Sweden
SME Energy CheckUp Project (2012) State of the art–retail
SUPER Smart Project (2016) Report 2–eco-friendly supermarkets-an overview
SUPER-HEERO Project (2021a) D2.2–Guidelines for the implementation and financing of energy efficiency measures in supermarkets
SUPER-HEERO Project (2021b) D2.1–Renovation measures catalogue for supermarkets
Van der Sluis SM (2017) Performance indicators for energy efficient supermarket buildings. In: 12th IEA heat pump conference

# Chapter 8
# Ideas for Improved Energy Saving Constructions for Windows

**Iris M. Reuther**

**Abstract** The adaptation of existing buildings is a key aspect of future-oriented building as utilising so-called grey energy for as long as possible contributes significantly to sustainable construction. This article considers why building envelopes are particularly important nowadays and focuses on the weakest point of a design that frequently occurs: the windows in punctuated facades of existing buildings. These often remain a weak point even following the energy-conserving renovation of a building, particularly the connection to the window reveal. This article demonstrates a major opportunity for improving this situation, namely by using a modern adaptation of the outer windows found in older buildings in certain regions of the German-speaking area. Precisely how this historical example can be adapted to modern window construction will be explained in principle and by example.

**Keywords** Building envelope · Energy saving · Outer windows

## 8.1 Introduction

### *8.1.1 A Worsening Situation and an Idea to Improve It*

The calls to save energy, known for years, have led to the construction of increasingly well-insulated buildings. In the interest of sustainability, the transmission heat losses of the building shell are also reduced in existing buildings by means of various measures. Where it was normal previously that thermal comfort was not generally restricted by structural or ventilation measures, this premise has now been abandoned. Already in autumn 2021, gas prices in Europe and the US rose to record levels (Finanzmarktwelt 2021). As a result of the conflict between Russia and Ukraine, which has also been fought out militarily since February 2022, energy

I. M. Reuther (✉)
Jade Hochschule Oldenburg, Ofener Straße 16/19, 26121 Oldenburg, Germany
e-mail: iris.reuther@jade-hs.de



P. Mendonça et al. (eds.), *Proceedings of 2023 International Conference on Green Building*, Environmental Science and Engineering,
https://doi.org/10.1007/978-3-031-43478-5_9

prices have risen in many European countries, especially those for gas (Agrar heute 2022). Although gas prices on the stock exchanges are now falling again (Agrar heute 2022) or (Council of the European Union 2023), the costs for many German consumers remain high. This has led and still leads to some consumers only heating their residential or business premises to lower indoor temperatures for cost reasons. For public buildings in Germany, a maximum indoor temperature of 19 °C has even been stipulated since September 2022. Areas in such buildings "where people are not permanently present may no longer be heated at all" (Die Bundesregierung 2023).

Lower indoor temperatures may make sense for energy policy or short-term economic reasons—from a structural physics point of view, they are fatal in many cases. The absorption capacity of water in the air depends on the temperature. Falling temperatures lead to a lower water vapour saturation pressure, and consequently often to condensation. It can be expected, especially in heavily used, heated interiors without artificial ventilation, that condensation will first form at the weak points of the building shell due to lower surface temperatures, followed soon afterwards by mould. While artificial ventilation for residential buildings is relatively common in Scandinavia and certain southern European regions have milder winters, the above-mentioned risk is increased in Germany in particular. This problem has existed for years due to airtight building shells, which is why the Federal Environment Agency, for example, published an extensive publication "Mould in indoor spaces" as early as 2003. In this publication, 10% of households were already found to have a problematic mould infestation (Umweltbundesamt 2003) p. 11. In 2017, a guide was published on, among other things, the prevention of mould infestation in buildings (Bundesumweltamt 2017), similar in GreenMatch (2023). It is certainly not without reason that an article published in 2012 by the Federal Environment Agency on proper ventilation was published in an updated form in November 2022 (Umweltbundesamt 2012). Conclusion: this long-standing problem is more topical than ever in the winter of 2022/23.

The author of this article has been working on building damage prevention for years (see e.g. (Reuther 2018, 2015)). The currently increased risk of mould in indoor spaces led to the reflections in this article. Even though it is not only the author who hopes for an early end to both the war and the "gas crisis" (Umweltbundesamt 2022) declared in Germany as a result of it, the results of this article are generally applicable as measures to reduce the aforementioned risk.

## 8.2 Ideas for Improved Energy Saving Constructions for Windows

### *8.2.1 Containment*

Mould that occurs indoors, as mentioned, develops on surfaces that are colder than others. This means that the lower interior temperature is not a problem (Pech and Pöhn 2018) p. 52, unless there are also design weak points (Bundesumweltamt 2017). In addition to geometric weak points such as external corners, material- or environment-related thermal bridges (Pech and Pöhn 2018) as well as structural faults, one component is regularly known as a potential weak point: the windows (Bundesumweltamt 2017) p. 53, (Pech and Pöhn 2018) p. 33, (GreenMatch 2023), or more precisely: the window frames and the adjoining reveal. This is shown by images taken with the thermal imaging camera, as exemplified by the figures in Fig. 8.1, which indicate high heat loss at the window frames even in well-insulated buildings built according to the latest standards (left). The isotherms in Fig. 8.2 also illustrate this. We aim to address a possible solution to this problem in the following.

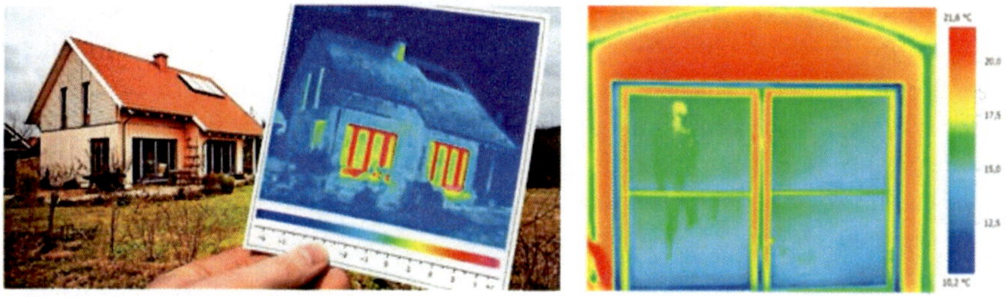

**Fig. 8.1** Examples of thermal bridges in windows, images taken using thermography from BauMentor (2023) (left) and Pech (2018) (right)

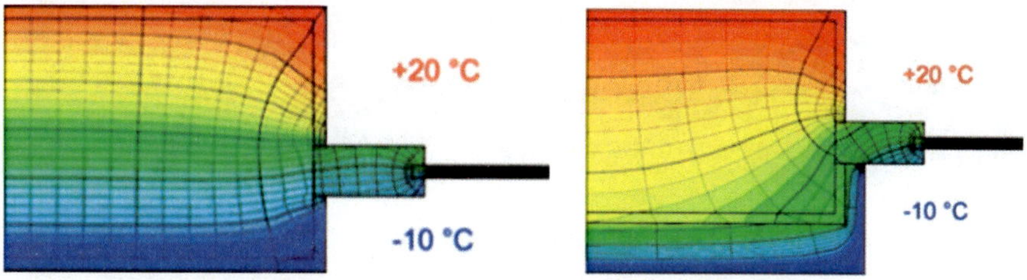


**Fig. 8.2** Examples of weak points on window frames, two variants from Pech and Pöhn (2018) p. 39: on the left without, on the right with external insulation of the wall, here also in reduced thickness introduced into the reveal

### 8.2.2 Current Approaches to Solutions

Only since glazing has become better and better in terms of thermal insulation has the edge seal become proportionally less favorable, and since the development of inert gas fillings in the intermediate pane frame, the window frame has regularly been the weakest point of windows in terms of thermal insulation (Pech et al. 2005). This area is less at risk of being colonised by mould due to the usual materials, but the adjacent reveal is increasingly at risk. Now it is not only theoretically possible to additionally insulate this frame component, which is especially possible in the case of plastic or wood/aluminum windows by insulating in hollow chambers. The additional, outside insulation of plastic window profiles can also be found (Pech et al. 2005) p. 51, in addition to this even DIY tips published in July 2022 by a public authority subsequently to stick strip EPS insulation to the profiles on the inside (LandesEnergieAgentur 2022). Expert architects and engineers understand that these measures tend to lead to an increased risk of mould. Unsurprisingly, the same public body posted a DIY tip for soffit insulation online in February 2023 (LandesEnergieAgentur 2023). However, this has several weaknesses: from the removal of the window strips, which are supposed to create the vapour-tight connection if necessary, the combustibility of the insulation material, possible cavities under the EPS boards, the insertion of liquid sealant in the window reveal incl. the resulting problems with recyclability, to the execution by laypersons and the question of why the more delicate interior insulation is proposed. Figure 8.3 shows a photo of the DIY instructions on the left and an isothermal curve on the window in the case of internal insulation (but not in the reveal) on the right. As positive as energy-saving measures are in principle, the author considers some of them to be potentially damaging; moreover, there is not always enough space in the window reveal for an interior insulation.

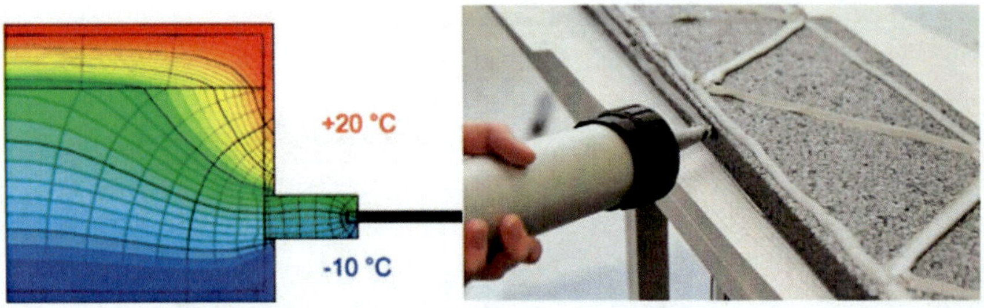


**Fig. 8.3** Photo from LandesEnergieAgentur (2023) with gluing instructions for EPF boards (left), plus another example of weak points on window frames from Pech and Pöhn (2018) p. 39 (right): here a reveal in a building with interior insulation

### *8.2.3 Historical Development*

This is why this chapter takes a look at historical window structures. Although these have not been displaced by modern variants without reason, they still have their advantages today. There were probably already windows with two glass levels in Roman times (Holzmanufaktur Rottweil 2020), but such designs later fell into oblivion. Even single-glazed windows were a sign of prosperity in the Middle Ages and the Renaissance. It was only from the seventeenth century onwards that so-called front windows, also known as winter windows or attachment windows, became known (Huckfeldt and Wenk 2009) p. 295. As an alternative to this element, which was usually mounted with fixed glass in front of the actual window during the cold season, there were also temporary windows on the inside, so-called winter casements, which could always be opened. Both designs reduced draughts through open joints and took advantage of the fact that stagnant air has a very good thermal insulating effect. Although the air was only stagnant to a limited extent due to a lack of sufficient seals, it nevertheless provided a significant improvement in thermal insulation. The so-called box window was developed from the latter window through the structural connection of both window levels. From the eighteenth century onwards these were installed in bourgeois and town houses throughout the German-speaking world, sometimes using slightly different names. (Huckfeldt and Wenk 2009) p. 296, (Holzmanufaktur Rottweil 2020) p. 13. Schematic drawings of these three types are attached in Fig. 8.4 as an illustration.

Box windows were very widespread, sometimes even prescribed by municipal ordinances (Holzmanufaktur Rottweil 2020). It was not until around 1920 that laminated windows became more widespread; followed much later by the insulating glass windows that are common today. As a result of the above, box-type windows became very widespread in German-speaking countries (Huckfeldt and Wenk 2009). In Austria alone, it is assumed that there are still 10 million box-type windows in existence (Ruisinger and Grobbauer 2012). In Germany, due to the massive destruction of the building stock during the Second World War, a renewed awareness of the value of the architectural heritage and the protection of historical monuments has led to more attention being paid to the remaining estimated 74 million box-type windows (Holzmanufaktur Rottweil 2020). As a result, there are now a number of publications on the energy-efficient renovation of such windows, which are

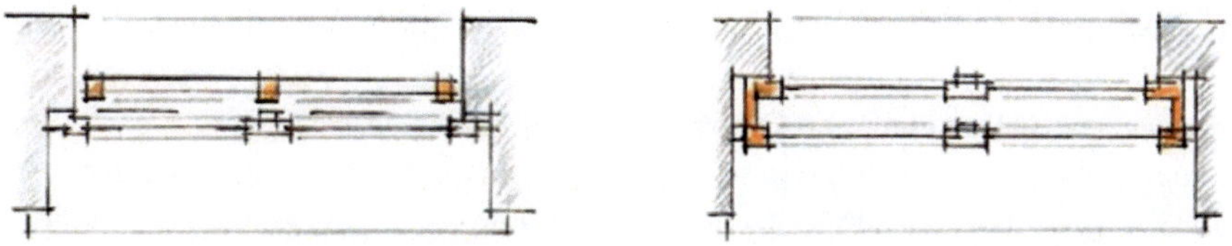

**Fig. 8.4** Front window (left) and box window (right) according to Huckfeldt and Wenk (2009) p. 296, horizontal sections, each own illustration

considered to be very durable due to their high quality (Holzmanufaktur Rottweil 2020) or (Historic England 2023). Years ago, the author was also involved in a project with such historic windows (Nickl Partner Architekten 2007), in which the inner sashes were fitted with seals as well as insulating glazing (similar to that described under (Holzmanufaktur Rottweil 2020; Historic England 2023; Kanton Bern 2023) or (Wisconsin Historical Society 2023)). This led to a considerable improvement in both airtightness and thermal insulation without any loss of heritage protection.

The essential feature of the box-type windows is the above-mentioned air layer between two window sash levels, which is advantageous in terms of thermal insulation. This principle was also discussed for large-scale façades as early as around 1850 and already served as a model for individual industrial buildings about 50 years later (Holzmanufaktur Rottweil 2020) p. 14, 15. Double-skin glass façades, mainly for office buildings, did not become common until the 1980s, because by then sun protection and air-conditioning had also been developed to a correspondingly high degree and the need for improved sound insulation could also be taken into account through these structures. It remains to be seen to what extent these glass double facades will actually be energetically useful and sustainable in the long term. In any case, even such ultra-modern office buildings make use of the aforementioned historical principle.

### *8.2.4 Building Physics Properties of Box-Type Windows or Single Glazing with Front Windows*

With regard to sound insulation, the window designs discussed in this chapter are advantageous (Holzmanufaktur Rottweil 2020; Berner Fachhochschule 2023). Air permeability, on the other hand, depends very much on the condition of the elements. In an energetically renovated state, air permeability classes of 2–4 according to EN 12207 are achieved, whereby 4 is the best achievable value (Berner Fachhochschule 2023) and class 2 is sufficient for buildings with up to two full storeys (Huckfeldt and Wenk 2009) p. 33. The airtightness of the building shell is considered a prerequisite for good thermal insulation (Pech and Pöhn 2018) p. 15, but this is hardly ever achieved, especially with the front windows that are usually installed without seals. However, individual research studies show interesting results here. For example, (Huckfeldt and Wenk 2009) p. 298 assumes a continuous flow through the space between the panes. Simulations and measurements show a rather still layer of air and, correspondingly, the thermography shows a kind of buffer zone between the two window levels (Ruisinger and Grobbauer 2012). The latter is shown in Fig. 8.5. It is even stated elsewhere that the low flow through the cavity "achieves a heat recovery effect", which reduces the transmission heat losses of box-type windows and the energetic difference to modern windows is reduced (Gronau and Helbig 1998) p. 78, similar described in Wisconsin Historical Society: Advantages of Maintaining Your Historic Windows (2023). Even for new buildings, excessively sealed connections

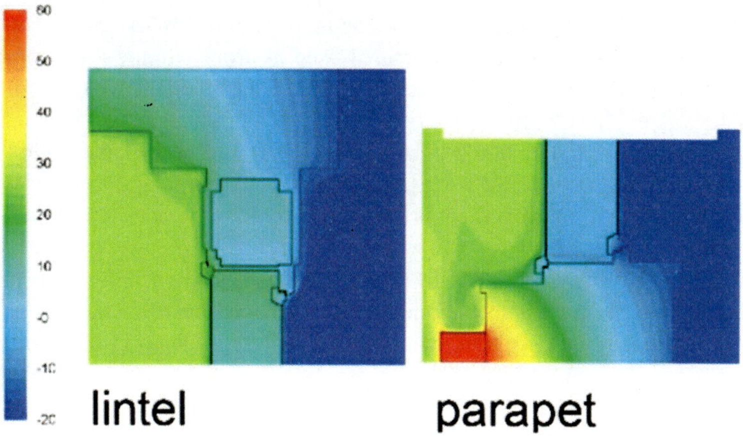


**Fig. 8.5** Temperature field of a box-type window in a divided vertical section (Ruisinger and Grobbauer 2012) p. 13, modified illustration (left: lintel, right: parapet)

are now being discouraged because this can lead to condensation and therefore mould in the interior. This is why "measures to increase air passage" are even advised ((Pech et al. 2005) p. 58 as well as (Holzmanufaktur Rottweil 2020) p. 7). From this, the author concludes that low air impermeability is not energetically disadvantageous.

In fact, it is concluded in Holzmanufaktur Rottweil: Das Fenster im 20. (2020), that "the energy savings demanded in the climate debate can be achieved particularly well with double-skin or multi-skin window constructions". In addition, box-type windows have an advantage in terms of building physics that is relevant to the problem described at the beginning: due to the "large structural depth, the thermal bridge effect with the consequence of very low surface temperatures is avoided" (Gronau and Helbig 1998) p. 75. The purely visual comparison between Fig. 8.2 and Fig. 8.5 clearly shows this. This shows that the box-type window is a structure that is even better than modern windows in terms of mould prevention.

### *8.2.5 Possible Transfer of the Benefits to Modern Windows*

At present, it is not a question of replacing modern insulated windows with replicas of historical windows. The effort for this alone would be enormous and not justifiable because of the grey energy required. Rather, this chapter will discuss the possibilities of transferring the benefits of the historic window structures to modern windows in order to improve them. The focus here is on reducing the increased risk of mould in window reveals, as described in Sect. 9.2.1. The simple approach is: greater building depth and therefore a stagnant air layer.

In specific terms, this could be achieved by means of front windows. Historical constructions, such as those secured with storm hooks and hung in front of the actual windows during the cold season, are analogous to Kanton Bern (2023) shown in Fig. 8.6. Now, openable window structures used as front windows are somewhat more complex to manufacture and the dead weight as a load on the window sill is not to be disregarded. This is why we are considering whether a hanging system including a distance profile for fixed, single-glazed windows would be a viable option. This idea is illustrated in Fig. 8.7. Both ideas should be able to be implemented for wooden windows. This type of improvement is also conceivable for window structures made of aluminium or steel.

However, very many windows in perforated façades, especially in residential construction, are plastic windows. Due to the lower load-bearing capacity of these elements or the load-bearing core, which lies at a certain depth of the profile, the

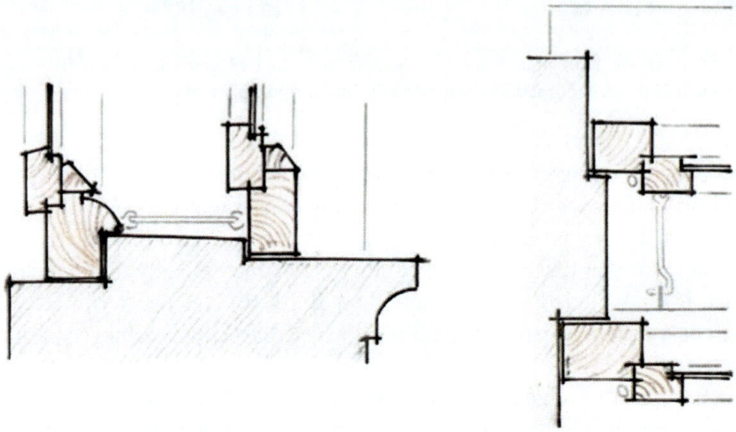

**Fig. 8.6** Front window in vertical (left) and horizontal (right) part-section according to Kanton Bern (2023), own illustration

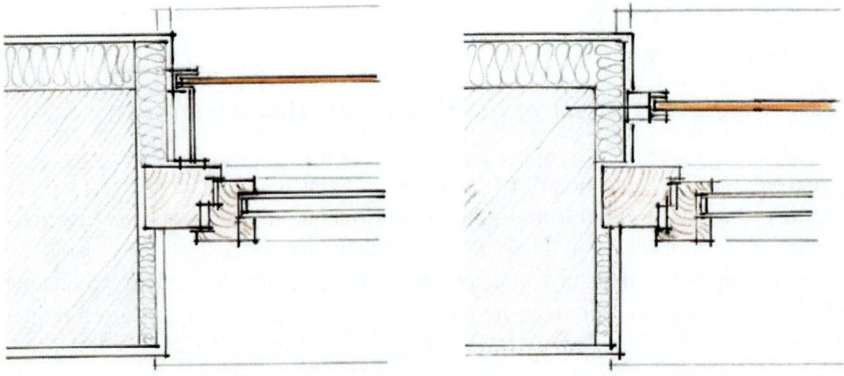

**Fig. 8.7** Two fundamentally conceivable options for front windows (orange coloured) in front of modern insulating glass windows, each horizontal detail and own illustration

author is rather critical of this application. These windows have a market share of almost 60% in Germany (Baulinks 2020). This is why we suggest as an alternative for these—if it is structurally available—namely to use the roller shutter rail as a support for single glazing and thereby achieve a buffer zone. In principle, roller shutters made of transparent plastic would also be conceivable, as it is well known that closed roller shutters contribute significantly to energy saving (Fensterbau Ratgeber 2021). Generally speaking, the guide rails or supports of sun protection elements could be used to serve as a substructure for front windows or even just glass panes during the cold season. These are also shown in principle in Fig 8.7. A hermetically sealed, stagnant air layer is certainly not achieved with these structures, but it is also not considered absolutely necessary.

However, there are two main points to clarify: the effort required to install and dismantle the front windows can be relatively large, depending on the type and height of the building, which is why they are considered suitable mainly for single-family houses and terraced houses. The second question regarding ventilation also needs to be clarified. In the case of new buildings, artificial ventilation may be assumed as a rule. Then it is not mandatory that the front windows can be opened. For older buildings, in the author's opinion, either compromises (i.e., not fitting all windows with front windows) or actually openable front windows, which are more expensive to manufacture, would make more sense. However, due to their own weight, these types of structures can hardly be attached to the guide rail of a sunshade.

In principle, therefore, although there are a few challenges to overcome, it is still conceivable in principle to implement the principle of box-type windows in the form of front windows as a way of improving modern windows. This could effectively counteract the increasing danger of mould formation in window reveals and also reduce heat loss from buildings. However, there is no doubt that both further theoretical research and practical studies are needed to implement this idea.

## References

Agrar heute (2022) Gaspreise gehen durch die Decke—das Schlimmste kommt noch. https://www.agrarheute.com/management/finanzen/gaspreise-gehen-decke-schlimmste-kommt-noch-596893. last accessed 06 02 2023

Agrar heute (2022) Gaspreise fallen um 30 Prozent—Verbraucher haben nichts davon https://www.agrarheute.com/markt/diesel/gaspreise-fallen-um-30-prozent-verbraucher-haben-nichts-davon-598036, last accessed 06 02 2023

Baulinks/BauSites GmbH (2021) Fenster- und Türenmarkt Deutschland 2020 und 2021 https://www.baulinks.de/webplugin/2021/0736.php4, last accessed 01 04 2023

BauMentor (2021)s Thermografie: Wärmeverluste erkennen und Kosten sparen https://baumentor.de/thermografie-waermeverluste-kosten-sparen/, last accessed 31 03 2023

Berner Fachhochschule (2023) Schallschutz und Luftdurchlässigkeit historischer Fenster https://www.bfh.ch/.documents/ris/2016-751.502.112/BFHID-1981910136-5/Broschuere-dt.pdf, last accessed 07 02 2023

Bundesumweltamt (2017) Leitfaden zur Vorbeugung, Erfassung und Sanierung von Schimmelbefall in Gebäuden https://www.umweltbundesamt.de/sites/default/files/medien/421/publikationen/uba_schimmelleitfaden_final_bf.pdf, last accessed 07 02 2023

Council of the European Union (2023) Infographic—A market mechanism to limit excessive gas price spikes https://www.consilium.europa.eu/en/infographics/a-market-mechanism-to-limit-excessive-gas-price-spikes/, last accessed 17 04 2023

Die Bundesregierung (2023) Energieversorgung sichern—Maßnahmen zum Energiesparen https://www.bundesregierung.de/breg-de/themen/klimaschutz/energiesparmass-nahmen-2078224, last accessed 06 02 2023

Fensterbau Ratgeber (2021) Rolladen Wärmeschutz https://www.fensterbau-ratgeber.de/fenster/sicht-und-sonnenschutz/rolladen/rolladen-waermeschutz/, last accessed 01 04 2023

Finanzmarktwelt (2021) Gaspreis in Europa auf Rekordhoch, in USA auf 7-Jahreshoch https://finanzmarktwelt.de/gaspreis-in-europa-auf-rekordhoch-in-usa-auf-7-jahreshoch-212518/, last accessed 06 02 2023

GreenMatch (2023) Black Mould on Windows: A Guide to Removal & Prevention https://www.greenmatch.co.uk/windows/moulding, last accessed 17 04 2023

Gronau J, Helbig S (1998) Das Kastenfenster—energetisch besser als sein Ruf! Bauphysik: Berichte aus Forschung und Praxis. pp 75–85

Historic England (2023) Modifying Historic Windows as Part of Retrofitting Energy-Saving Measures https://historicengland.org.uk/advice/technical-advice/retrofit-and-energy-efficiency-in-historic-buildings/modifying-historic-windows-as-part-of-retrofitting-energy-saving-measures/, last accessed 17 04 2023

Holzmanufaktur Rottweil (2020) Das Fenster im 20. Jahrhundert https://www.holzmanufaktur-rottweil.de/fileadmin/user_upload/Publikationen/PDF/Kastenfenster_Inhalt_low.pdf, last accessed 01 04 2023

Huckfeldt T, Wenk H-J et al (2009) Holzfenster—Konstruktion. Sanierung, Wartung, Rudolf Müller Verlag, Köln, Schäden

Kanton Bern (2023) Beispiele für Fenstersanierungen https://www.kultur.bkd.be.ch/content/dam/kultur_bkd/dokumente-bilder/de/themen/denkmalpflege/bauvorhaben-und-planungen/umbauen-und-restaurieren/fenstersanierungen/Beispiele-Fenstersanierungen.pdf, last accessed 07 02 2023

LEA LandesEnergieAgentur Hessen (2022) Fensterrahmen dämmen https://redaktion.hessen-agentur.de/publication/2022/3852_DIY-Anleitung-Fensterrahmendaemmen.pdf, last accessed 31 03 2023

LEA LandesEnergieAgentur Hessen (2023) Fensterlaibungen innen dämmen https://redaktion.hessen-agentur.de/publication/2023/3996_DIY-Anleitung-Laibungsdmmung.pdf, last accessed 01 04 2023

Nickl & Partner Architekten (2007) Denkmalgeschützte Pavillons Haus 6 und Haus 2 des Klinikums München—Schwabing. https://www.nickl-partner.com/projekte/healing-architecture/denkmalgeschuetzte-pavillons-haus-6-und-haus-2-klinikum-muenchen-schwabing/, last accessed 01 04 2023

Pech A, Pöhn C (2018) Baukonstruktionen—Band 1: Bauphysik. Birkhäuser Verlag, Wien

Pech A, Pommer G, Zeininger J (2005) Baukonstruktionen—Band 11: Fenster, Springer Verlag, Wien

Reuther IM (2015) Acceptance by durability—Quality assurance and insurance sustain innovative facades. In: Advanced Building Skin, 10th Conference on Advanced Building Skins, pp 865–870, Bern

Reuther IM (2018) Best Practice Gesucht—Strukturen baulichen Qualitätsmanagements im Hochschulneubau anhand exemplarischer Projekte in Österreich. Technische Universität Graz, Graz, Deutschland und Luxemburg

Ruisinger U, Grobbauer M (2012) Innendämmung, Holzbalkenköpfe und Kastenfenster in der Sanierung. In: Bauphysiktagung 2012, TU Graz, Graz

Umweltbundesamt (2003) Schimmelpilze in Innenräumen https://www.umweltbundesamt.de/publikationen/schimmelpilze-in-innenraeumen, last accessed 07 02 2023

Umweltbundesamt (2012/2022) Wie lüfte ich richtig?—Tipps und Tricks zu Schimmelvermeidung https://www.umweltbundesamt.de/themen/gesundheit/umwelteinfluesse-auf-den-menschen/schimmel/wie-luefte-ich-richtig-tipps-tricks-zu, last accessed 07 02 2023

Umweltbundesamt (2022) Mit Klimaschutz durch die Gaskrise https://www.umweltbundesamt.de/sites/default/files/medien/1410/publikationen/2022-10-26_texte_111-2022_gas_wasserstoff_und_klimaschutz_utf_bf.pdf, last accessed 07 02 2023

Wisconsin Historical Society (2023) Advantages of Maintaining Your Historic Windows. https://www.wisconsinhistory.org/Records/Article/CS4302, last accessed 17 04 2023

# Chapter 9
# Construction Standard and Input–output Analysis of Green Low-Carbon District of Enterprise A

**Zhuangzhuang Li, Na Zheng, Yue Dai, Lihong Zheng, Kexin Wu, and Yongqiang Kong**

**Abstract** The low-carbon development of urban construction is an important part of green low-carbon transformation for society. Based on the business needs of comprehensive urban development in enterprise A, this study integrated advanced technologies and development hotspots at home and abroad, explored the construction content of green low-carbon district suitable for the enterprise's business. A development case was selected as the analysis model. The incremental construction cost generated by the development according to the standard was analyzed by means of inquiry, online search, reference to the national price limit and analogy. By evaluating the incremental cost of technology and investment benefits, the feasibility of standard implementation was weighed. The incremental cost of building green low-carbon district of different levels ranged from 132.9 to 1393.3 thousand yuan/hm$^2$. After fully completion and operation, the reduction of $CO_2$ emission can reach 27.5–58.5%. The input–output benefits confirmed its feasibility.

**Keywords** Green low-carbon district · Enterprise standard · Incremental cost · Investment benefit

## 9.1 Introduction

In September 2020, General Secretary Xi Jinping made a solemn commitment to the world at the United Nations General Assembly that "carbon peak by 2030 and carbon neutral by 2060", putting forward higher requirement for China's low-carbon development. According to the data of the seventh national census, the proportion of

---

Z. Li (✉) · Y. Dai · K. Wu · Y. Kong
China Construction FangCheng Inverstment & Development Group Co. Ltd., Beijing, China
e-mail: zzlee.2009@163.com

N. Zheng · L. Zheng
Tianjin Eco-City Green Building Research Institute, The Sino-Singapore Tianjin Eco-City, China



P. Mendonça et al. (eds.), *Proceedings of 2023 International Conference on Green Building*, Environmental Science and Engineering,
https://doi.org/10.1007/978-3-031-43478-5_10

urban population in China has reached 63.9% by 2020, with an increase of 14.2% over 2010 (Chewei and Yifei 2021). The green and low-carbon development in urban areas plays vital roles in reducing social carbon emission, and green low-carbon district emerges as The Times require (Hongtao and Juanjuan 2018). Green low-carbon district was the urban construction area that is planned, constructed and operated in accordance with the requirements of resource saving and environmental friendliness in terms of spatial layout, infrastructure, architecture, transportation, ecology, green space, industry, etc.

Enterprise A is a professional platform that actively participates in and leads the new urbanization construction business in China. It specialized in the comprehensive development of urban land and urban operation business, and has accumulated rich urban construction capacity and experience. Driven by the concept of green and low carbon, the enterprise ushered in a new stage of innovation and breakthrough. Based on the construction needs of green low-carbon district and the opportunity of enterprise business expansion, this study integrated advanced technologies and development hotspots at home and abroad, explored the construction content of green low-carbon district suitable for the enterprise's business, formulated the construction standard system with different development depths, and weighed the feasibility of standard implementation by analyzing the incremental cost and investment benefit of key technologies.

## 9.2 Construction Standard of Green Low-Carbon District for Enterprise A

The establishment of the construction standard for green low-carbon district was based on the technical requirements of <Evaluation standard for green eco-city> GB/T 51,255. In order to meet the business needs of comprehensive urban development in enterprise A, the standard integrated advanced technologies and development hotspots at home and abroad, comprehensively considered the practicality and economy of technology, and formulated the construction standard system with different development depths (Hongbo and Zhongli 2020; Shaoxiong and Xiaoyu 2020; Xiaoyu and Qihang 2020; Yijiang and Peng 2021).

The standard adhered to the principle of adaptation to local conditions. After comparing high-quality green low-carbon district at home and abroad, the construction standard including basic level, upgraded level and exemplary level were formed (Tan and Yang, 2017; Lou and Jayantha 2019). Taking the six core idea (top-level design, energy saving and low-carbon, green integration, comfort and livability, coordinated development of industry and scientific operation) as starting points, we formulated technology paths that were achievable and easy to operate, providing development ideas for projects (Hui 2020).

The basic level focused on the optimization of ecological environment, green transportation and humanistic environment, so as to build a beautiful living environment. The upgraded level optimized the land use mode and transportation system on the basis of the basic level. Meanwhile, it strengthened the governance of ecological environment protection and energy saving and carbon reduction, and improved the management level of buildings and city by building high-star green buildings and intelligent platform. The content of the exemplary level comprehensively covered the fields of ecological protection, environment governance, transportation, green buildings, resource utilization, carbon emission management, information-based management, industrial guidance, etc., representing the most advanced urban development level in China.

## 9.3 The Estimation of Incremental Cost

According to the construction standard for enterprise A, a development case was selected as the analysis model, which of the total land area is about 1000 $hm^2$, the total construction area is 8 million m2, and the investment scale is 12 billion yuan. The incremental construction cost generated by the development according to the standard was analyzed by means of inquiry, online search, reference to the national price limit and analogy (Zuda 2012). The analysis results were shown in Table 9.1.

With the improvement of development level, the technologies of basic-level green low-carbon district were basically popularized, as a result, the incremental cost is only 132.9–181.7 thousand yuan/$hm^2$. The incremental cost of high-level green low-carbon district ranged from 651.1 to 1393.3 thousand yuan/$hm^2$, accounting for 1.6% to 3.5% of the total investment and 1.1–2.4% of the land transfer income for this case, indicating that the background condition for urban construction in China was better.

The incremental cost of upgraded-level green low-carbon district accounted for 1.6–2.2% of the total investment and about 1.1–1.5% of the land transfer income, which was a relatively acceptable level. Taking the upgraded level as an example, the distribution of increment cost was analyzed. The mean values were adopted. The distribution of incremental cost by category was presented in Fig. 9.1.

**Table 9.1** Analysis results of incremental cost for green low-carbon district construction

| Levels | Total increment cost (ten thousand yuan) | Proportion of incremental cost in total investment (%) | Incremental cost per unit area (thousand yuan/ $hm^2$) | Proportion of incremental cost in land transfer income (%) |
|---|---|---|---|---|
| Basic level | 3988–5450 | 0.3–0.5 | 132.9–181.7 | 0.2–0.3 |
| Upgraded level | 19,533–26,682 | 1.6–2.2 | 651.1–889.4 | 1.1–1.5 |
| Exemplary level | 31,603–41,800 | 2.6–3.5 | 1053.4–1393.3 | 1.8–2.4 |

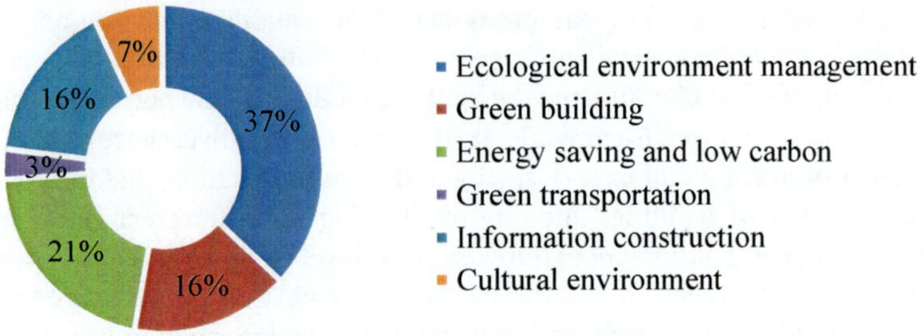


**Fig. 9.1** Incremental cost distribution chart of upgraded-level green low-carbon district

The incremental cost of land use and industrial system establishment can be ignored. The incremental cost of ecological environment governance, energy saving and green buildings accounted for the highest proportion, accounting for 37, 21 and 16%, respectively. Information construction was attached great importance with the proportion of 16%. The creation of humanistic environment mainly depended on personnel organization and resource linkage, which can obtain a good humanistic environment with less investment. As for the transportation system, in terms of China's current development level, it has basically met the requirements of green low-carbon district, with the incremental cost of 3%.

## 9.4 The Prediction of Investment Benefits

The advantages of green low-carbon district were embodied in the aspects of environmental friendliness, reduction of carbon emission, economic progress and life quality improvement, which have good environmental, economic and social benefits.

### *9.4.1 Environmental Benefit*

Carbon emission control was of great significance to environmental protection (Chamaratana and Knippenberg 2020). The case for incremental cost analysis was used as the model to analyze the emission reduction (Yang and Feng 2014). This study analyzed the energy-saving effect or carbon emission characteristics of each unit, calculated the carbon emission reduction of each unit through activity volume, energy consumption reduction level and emission factors. Then the total added the amount of carbon removed by carbon sink and the amount of renewable energy replaced, obtaining the total carbon emission reduction (Zuda and Jingyi 2016). The model formula was as follows.

$$C = \sum_{i} A_{di} \times e_i + W \times e_w + A_g \times e_g + C_{re} \tag{9.1}$$

where:

- C total annual carbon emission reduction in the detailed planning of green low-carbon district, $tCO_2e$/year;
- $A_{di}$ energy consumption reduction of class i carbon emission unit (i = 1,..., 4,6; 1 = building, 2 = transportation, 3 = industry, 4 = water resources, 6 = municipal facilities);
- $e_i$ energy consumption and emission coefficient of class i carbon emission unit;
- W waste recycling and resource treatment capacity, $m^3$;
- $e_w$ carbon emission coefficient of waste landfill, $tCO_2e/m^3$ a;
- $A_g$ urban green space, $hm^2$;
- $e_g$ carbon removal coefficient of green space, $tCO_2e/hm^2$ a;
- $C_{re}$ carbon emission reduction of renewable energy replacing conventional energy, $tCO_2e$/a.

The values of activities and reference factor of each emission unit in formula (9.1) referred to the research results of Ye's team (Zuda 2016). The evaluation results of carbon emissions were shown in Table 9.2. After fully completed and put into use, the green low-carbon district in accordance with the construction standard formulated by this research can reduce carbon emission by 240,300 $tCO_2e$/a, 403,600 $tCO_2e$/a and 511,700 $tCO_2e$/a in the basic, upgraded and exemplary level. Compared with the urban areas constructed under conventional development plans, the reduction proportion of carbon emission has reached 27.5, 46.2 and 58.5%, respectively, with significant carbon emission reduction effect.

### *9.4.2 Economic Benefit*

To build a green, ecological, livable and modern low-carbon district could enhance the regional value, attract developers, revitalize land resources and bring good economic returns. A Public–Private-Partnership project of the enterprise A was located far away from the urban core area. After being evaluated as a green low-carbon district, the land transfer price and commercial housing price are basically equal to the core area.

In addition, green low-carbon districts had a significant effect on reducing $CO_2$ emission. Carbon emission can be reduced by 240,300–511,700 $tCO_2e$/a. At present, the carbon trading markets are gradually opening up. In the future, urban areas can be considered as a unit for carbon trading, which have extremely high efficiency and kinetic energy incentives for low-carbon development.

For enterprises, to provide high-quality and professional services for the government, the income return tended to be more stable. The value of development project

**Table 9.2** Evaluation results of reducing carbon emission in green low-carbon district

| Carbon emission assessment parameters (ten thousand $tCO_2e/a$) | | Basic level | Upgraded level | Exemplary level |
|---|---|---|---|---|
| Emission reduction effect with green and low-carbon measures | Buildings | 13.65 | 24.86 | 35.79 |
| | Traffic | 4.79 | 7.53 | 7.98 |
| | Industry | / | / | / |
| | Water resource | 0.87 | 1.41 | 1.88 |
| | Waste | 0.23 | 2.14 | 2.14 |
| | Municipal infrastructure | 0.12 | 0.57 | 0.57 |
| | Carbon sink | 0.07 | 0.07 | 0.07 |
| | Renewable energy | 5.15 | 3.78 | 4.04 |
| | Total emission reduction of carbon | 24.03 | 40.36 | 51.17 |
| Total actual carbon emissions[a] | | 68.61 | 50.40 | 40.36 |
| Net carbon emissions[b] | | 63.39 | 47.06 | 36.25 |
| Total carbon emissions of conventional development schemes | | 87.42 | 87.42 | 87.42 |
| Carbon emission reduction ratio | | 27.5% | 46.2% | 58.5% |

[a]The evaluation results were based on the selected model parameters, and the land scale is about 1000 $hm^2$

[b]The total net carbon emission is the carbon emission after considering carbon sink and renewable energy

would be further improved compared with the ordinary area, which was helpful to obtain higher economic benefits.

### 9.4.3 Social Benefit

Under the social background of the urgent implementation of "carbon peak and carbon neutralization", practicing the concept of green development was an effective way to actively explore the way of low-carbon development. The positioning of the urban area could been improved by means of scientific planning, low-carbon construction, elaborative management and other effective measures. It was beneficial to improve the city cultural taste and create a green and low-carbon city card, thus enhancing social influence.

Enterprises A payed more attention to national policies, always practiced the concept of green and low-carbon development in new urbanization construction business, which was committed to meeting the needs of the government and society. In recent years, the government's recognition has been greatly improved. During this period, it has established a green and low-carbon brand image. Furthermore, the position in the industry has been further improved.

## 9.5 Conclusions

This study took the comprehensive urban development business as the starting point and formulated the construction standard of green low-carbon district applicable to enterprises A. The standard divided the development level of green low-carbon district into basic level, upgraded level and exemplary level, so as to match the optimal scheme according to the current situation of the district to be developed. In order to evaluate the practicality of the construction standard, this study comprehensively analyzed the incremental cost and investment benefits of green low-carbon district. The incremental costs of green low-carbon district constructed according to the three-level standards were 132.9–181.7 thousand yuan/$hm^2$, 651.1–889.4 thousand yuan/$hm^2$ and 1053.4–1393.4 thousand yuan/$hm^2$, respectively. After being fully completed and put into use, $CO_2$ emissions could be reduced by 27.5, 46.2 and 58.5%. In addition to gaining government recognition and enhancing social influence, it also enhanced the core competitiveness.

## References

Chamaratana T, Knippenberg L (2020) toward a low carbon city: community networks for developing and promoting carbon emission reduction behavior, Khon Kaen, Northeast Thailand. Int J Sustain Policy Pract 16(2):1–13

Chewei Z, Yifei C (2021) Long-term trends and implications of population change from the seventh census data (in Chinese). Xinhua Digest (15):3

Hongbo L, Zhongli M (2020) Discussion on the planning and design of sponge city construction in green eco-district (in Chinese). Urban and Rural Studies 000(005):36

Hongtao L, Juanjuan L (2018) The impact of urbanization on carbon dioxide emissions in China based on Input-output locally closed model (in Chinese). Sino-Global Energy 23(4):10

Hui H (2020) Comparative study on green eco-district planning and practice (in Chinese). Environ Dev 32(1):201–203

Lou Y, Jayantha WM (2019) The application of Low-carbon City (LCC) indicators—a comparison between academia and practice. Sustain Cities Soc 51:101677

Shaoxiong L, Xiaoyu L (2020) Practice of planning and design of green eco-district—Taking Zhangzhou west lake eco-park as an example to establish a national three-star green eco-district (in Chinese). Architecture & Culture 07:194–196

Tan, S., Yang, J.: A Holistic Low Carbon City Indicator Framework for Sustainable Development. Applied Energy, 185(pt.2): 1919–1930 (2017).

Xiaoyu L, Qihang W (2020) Green eco-district planning and design: a case study of Zhangzhou West Lake Eco-park (in Chinese). Design Community 04:58–64

Yang Z, Feng D (2014) Study on optimization model of total carbon emission control in urban area under uncertain conditions (in Chinese). Renew Energy Resour 032(012):1922–1927

Yijiang L, Peng W (2021) Application of green transportation concept in traffic planning of eco-city (in Chinese). Research 2015–30:285–285

Zuda Y (2012) Cost-benefit Analysis of controlled detailed planning for low carbon ecology (in Chinese). Urban Dev Stud 19(01):58–65

Zuda Y (2016) Carbon emission assessment model of the decision-making tool for control detailed planning of green eco-district (in Chinese). Urban Dev Stud 23(03):76–86

Zuda Y, Jingyi W (2016) Carbon emission assessment model of low-carbon planning and construction management in green eco-district (in Chinese). Modern Urban Res 31(02):84–92

# Chapter 10
# Research on the Innovative Management Model of a Company's Green and Low-Carbon Urban Construction

**Zhuangzhuang Li, Na Zheng, Yue Dai, Lihong Zheng, Weichang Cao, Guoyong Feng, Juanjuan Bao, and Yong Su**

**Abstract** Comprehensively analyzing the key points and difficulties in promoting of PPP construction projects for green and low-carbon urban, the new way for enterprise development was explored. By analyzing the drawbacks of the green low-carbon thinking in existing project management methods, aiming at the implementation of green technology system and carbon reduction measures, we have created an innovated management models for management optimization. The green low-carbon content management and control links were increased and coordination and cooperation among various departments was promoted. Finally, a management system for the implementation of work was established to provide institutional support.

**Keywords** Green and low-carbon urban area · Innovation management · Carbon emissions · Management model

## 10.1 Preface

Urbanization is an effective way to reduce the gap between urban and rural areas and the gap between rich and poor, and is an important symbol of national modernization. Since the reform and opening up, China's spatial urbanization rate has increased by 45.99%. However, the urban working population lacks the ability to purchase houses, and the level of population urbanization and spatial urbanization is not coordinated (Huan and Hongbing 2021). Since 2014, Public–Private-Partnership mode (PPP) has been promoted in our country (Changhui and Lei 2016). Operation mode of PPP Projects was shown in Fig. 10.1 (Fuya and Jiaxin 2021). In this development

Z. Li (✉) · Y. Dai · W. Cao · G. Feng · J. Bao · Y. Su
China Construction Fangcheng Inverstment and Development Group Co., Ltd., Beijing, China
e-mail: zzlee.2009@163.com; lncs@springer.com

N. Zheng · L. Zheng
Tianjin Eco-City Green Building Research Institute, Tianjin, China



P. Mendonça et al. (eds.), *Proceedings of 2023 International Conference on Green Building*, Environmental Science and Engineering,
https://doi.org/10.1007/978-3-031-43478-5_11

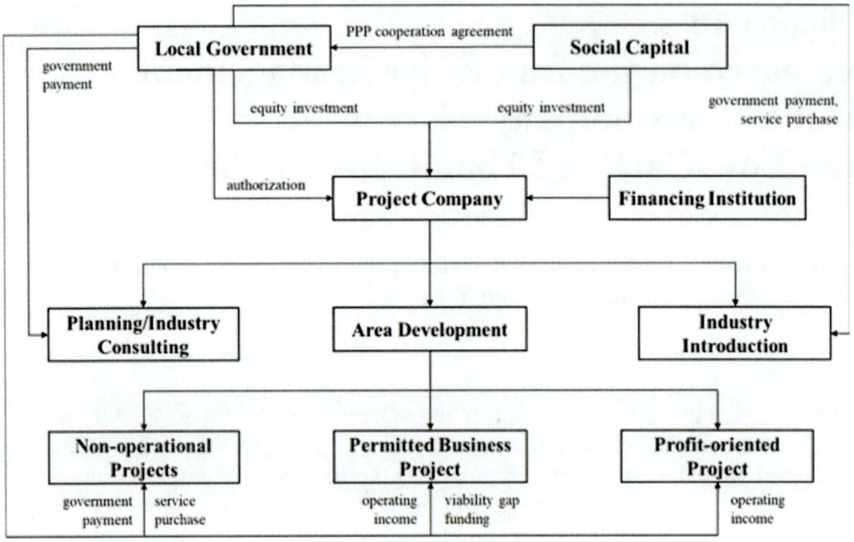


**Fig. 10.1** Operation mode of urban integrated development PPP projects

mode, the management thinking of green and low-carbon construction and operation regulates the balance between the high-quality development goal of the city and the construction of ecological civilization. Integrating the sustainable development thinking into the whole process of urban planning, construction and management, and implementing high-quality construction with urban areas as the main body of development is of great value to ecological environmental protection, optimization of energy structure and implementation of carbon emission management, and will comprehensively promote the green, low-carbon, sustainable and high-quality development of modern cities (Zifang 2018).

Building a green and low-carbon urban landscape requires more technical and cost inputs. However, on the premise of mature technology and controllable costs, green and low-carbon urban areas have not entered a vigorous development situation, which is due to the following reasons:

(1) The green low-carbon thinking needs to be implemented: the government and social capital have a low understanding of green and low carbon and lack substantive understanding, and the specific implementation and implementation are still not clear enough;
(2) Lack of guidance policies: The guidance and incentive system for the construction of green and low-carbon urban areas at the national and local levels still needs to be improved, and the construction enthusiasm of local governments is not high, lacking development momentum;
(3) Slow investment benefits: The incremental cost of high-quality urban construction is controllable, and the long-term benefits are endless. However, developers and users have limited awareness and are reluctant to pay for it.

In summary, while complying with social development trends, the company is also facing a series of tests brought about by the innovation of urban development thinking. Therefore, under the premise that the technical implementation is guaranteed, to further optimize the benefits of green and low-carbon urban construction, it is important to carry out research on the optimization of management mode based on the model of business subjects. Based on the comprehensive development of cities and towns, this study analyzes the current management situation of internal and external coordination and internal interaction within enterprises from the perspective of investment management, explores the working mechanism and management mode in the low-carbon city construction, optimizes management methods, and proposes driving measures, management methods, and guarantee mechanisms for advanced development content; To guide the government's development and investment thinking upward is conducive to improving the government's recognition of the construction of green and low-carbon urban areas and the incremental costs they bring, and to standardize the implementation of project construction downward to achieve the purpose of controlling costs and improving the quality of project construction. Emphasis is placed on promoting the implementation of low-carbon city construction content and improving the development quality.

## 10.2 Analysis of the Current Management Situation of the Company

The company focuses on the comprehensive development and urban operation of urban tracts of land, focusing on three business areas: old city renewal, new city development, and new town construction. The company adheres to the development thinking of "green and intelligent" and has been intensively applied in project implementation. The existing work mode is applicable to conventional urban comprehensive development projects, and the construction of low-carbon city areas lacks driving force and executive force (Mei and Xianjin 2021). Based on the project experience of conventional urban areas being developed by the company, the management disadvantages unfavorable to low-carbon city areas construction are analyzed and summarized as follows (Taohong and Zhe 2022). The management disadvantages mainly includes eight contents, namely: "Three level architecture" management system still needs to be improved, different understanding levels of green thinking, low-carbon internal control standards need to be improved, project quality control at the front end of investment needs to be strengthened, cost control system to be improved (Hua 2016; Danyang 2014), strengthen the cultivation of low-carbon talents, the establishment of project management database still needs to be improved, and the level of industry introduction needs to be improved urgently.

## 10.3 Research on Management Mode Under the Demand of Green Low-Carbon Construction

The construction content of urban comprehensive development projects is complex and the construction period is long, which requires strict organizational management to restrict project implementation. After years of exploration and practice, the company's business capabilities and organizational level have tended to improve. However, compared to conventional urban development, the construction of green and low-carbon urban areas highlights the development thinking of "green" and "low-carbon", and the application of relevant technical systems is more in-depth from early planning, design, construction, and operation (Yuan 2019). In order to drive the implementation of low-carbon strategies to ensure the quality of development, the company's overall management model needs to make corresponding changes with the innovation of construction content. Based on this business background, the green and low-carbon management mode was formulated in this study. The structure chart was presented in Fig. 10.2.

### *10.3.1 Planning Guidance System*

The top-level thinking is the decisive factor leading the development direction of the enterprise. When conducting enterprise planning research, the ideology of relevant departments must stand at the commanding heights of enterprise development, constantly absorb advanced technical theories and experience in the domestic and overseas, deeply understand the connotation of the low-carbon urban construction model, and integrate the advanced construction content and low-carbon thinking into the company's development strategy through enterprise development planning work (Wu 2015).

### *10.3.2 Cooperative Support System*

The top-level thinking is the decisive factor leading the development direction of the enterprise. When conducting enterprise planning research, the ideology of relevant departments must stand at the commanding heights of enterprise development, constantly absorb advanced technical theories and construction experience in the domestic and overseas, deeply understand the connotation of the green and low-carbon urban construction model, and integrate the low-carbon thinking into the company's development strategy through enterprise development planning work (Wu 2015).

**Investment Front End**. As the project controller of the front end of investment, investment expansion and investment management departments need to deeply

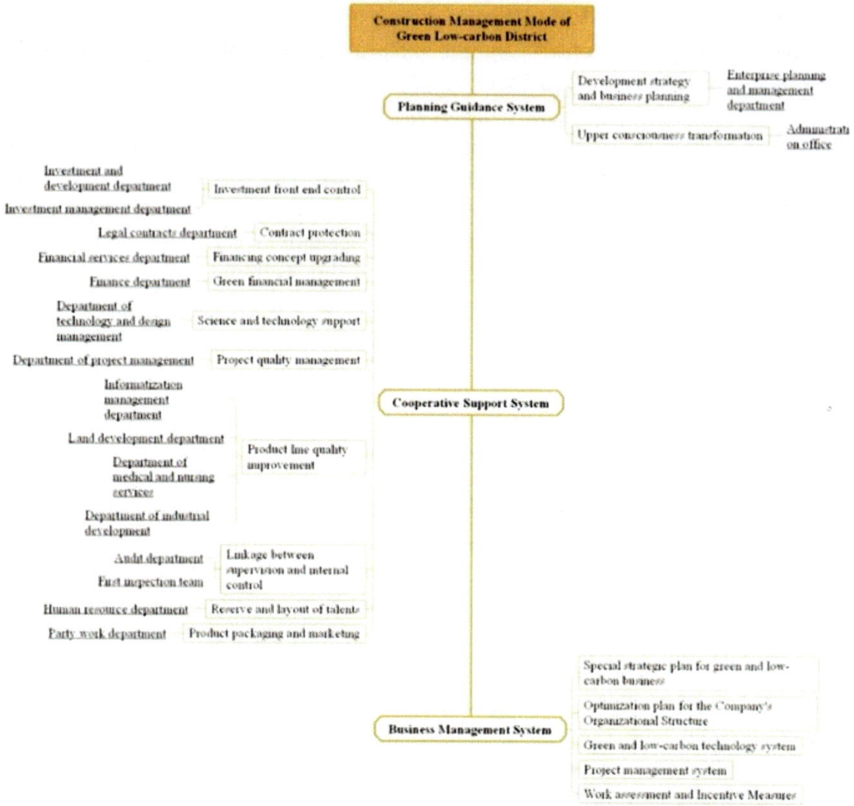


**Fig. 10.2** Green and low-carbon urban construction management mode structure chart

explore government needs and urban development positioning, find service entry points, deeply interpret national green and sustainable policies for government, then penetrate the construction thinking of carbon emission reduction. During the stage for organizing the feasibility study, project initiation, and review of the project, the construction requirements and content about low-carbon construction has been added, plan's feasibility is thoroughly studied, and strict checks are made through the review, actively controlling the development direction and quality of the project from the front end of investment.

After determining a green and low-carbon development plan, the human resources department should quickly respond to the company's strategic decisions, timely deploy talent reserve plans, cultivate or introduce talents with professional skills in green buildings, renewable energy utilization, sponge cities, carbon emission management, and provide backup forces for the implementation of the company's new development strategy.

**Investment Midrange**. The middle end of investment mainly involves comprehensive development project contract signing and project financing activities. Before the development of an area, the company signs a contract with the government. When drafting and reviewing the contract, the legal contract related departments should add information related to the construction content of the green and low-carbon urban area, ensuring the implementation from the legal level.

In the work deployment, financial business related departments should learn green low-carbon development thinking. During financing activities, it is useful to integrate the thinking of green and low-carbon, highlight the highlights of green and low-carbon urban development in bank loans. This action can not only enhance the project positioning height, but also help solve the financing difficulties, so that development activities can be funded.

**Investment Backend**. After the investment and construction of the project, the financial department will coordinate the use of funds, and be responsible for the planning and design, project progress, informatization, industrial development, and other management related departments to perform their respective management functions.

In terms of planning and design management, the content of green and low-carbon construction should be included in the regulatory scope, and detailed evaluation standards and progress assessment standards should be formulated. During the scheme review, a special review team for corporate strategies will be established for ensuring application of green low-carbon urban construction standards in the top-level design. At the same time, it is necessary to prevent excessive planning and design functions and overall control the balance between construction demand and cost.

When implementing supervision, the project management department cooperates with the planning and design department and the project company to incorporate continuable construction content in the range of supervision and develop detailed progress assessment standards and quality evaluation standards.

Green and low-carbon urban areas focus on information management and scientific operation. Information management departments should start investigating and studying the technical conditions, development costs, and solutions of the energy and carbon emissions information platform, develop implementation plans, and strive to promote the timely implementation of carbon emissions management (Zhaoyu 2021).

The industrial structure affects the carbon emissions during the operation stage. In the industrial planning stage, it is recommended that the industrial development department propose clear indicator requirements for the project company and formulate an industrial optimization plan (Jiawei 2020). Explore the potential of industrial park projects in urban areas, guide the entry of industries related to green buildings and green low-carbon urban areas. The domestic carbon trading market is becoming increasingly perfect. In the future, it can be considered to conduct carbon trading in urban areas as a unit, seeking benefits for government through low-carbon management methods, which has high efficiency and momentum incentives.

### *10.3.3 Business Management System*

The development of green and low-carbon urban areas has a long construction period and complex content. After the implementation of basic responsibilities, corresponding systems and guarantee systems will be established to standardize the work content through scientific and meticulous management. Through professional management, it ensures the formation of low-carbon business chain, technology implementation, as for as the improvement of management functions, providing a strong institutional guarantee for business development.

**Special Strategic Plan for Green and Low-Carbon Business**. Driven by the goal of "carbon peak, carbon neutral", low-carbon pathway is bound to become a mainstream trend in the future. Exploring the development direction and path of the enterprise, formulating special plans for strategic development, clarifying the business direction and implementing key points of low-carbon strategies will be the an important focus for future work.

**Optimization Plan for the Company's Organizational Structure**. The emergence of a new business system is accompanied by deepening work complexity, which will correspondingly lead to changes in the management structure. In the research of organizational structure, it is necessary to arrange ahead of time, innovate and optimize to lay a solid foundation for the expansion of new businesses.

**Green Low-Carbon Technology System**. Referring to low-carbon positioning of urban construction, starting from ecological environment, municipal facilities, green buildings, energy and resource utilization, carbon emission management, information construction, etc., specify construction content and technical requirements, determine the company's technical capabilities, establish the low-carbon technology application system, and associate the technology library to ensure technical support compliance.

**Project Management System**. The green low-carbon application is complex, and existing management personnel have certain difficulties in mastering and managing these strategies. While inputting professional management talents, a complete set of management methods and systems should be established to guide the practice of low-carbon strategies throughout the entire life cycle of planning, design, construction, and operation, supervise and restrict the implementation of all aspects of work, and provide strong institutional protection for project construction.

**Work Assessment and Incentive Measures**. In order to coordinate the future development of green and low-carbon urban business, relevant departments such as planning should respond to development needs in a timely manner, adding assessment methods and standards for construction of urban areas to the existing assessment system, or using this content as an independent chapter for special review and assessment. At the same time, the incentive system of the company has been increased to provide incentives to groups or individuals who have performed outstanding work related to green and low-carbon urban projects, to stimulate employees' enthusiasm for learning and implementation (Chenglong 2019).

## 10.4 Conclusion

In the "carbon peak, carbon neutral" action, green low-carbon transformation and sustainable development has become the only way to high-quality development of enterprises. Whether an enterprise can shoulder the responsibility of green low-carbon attempt became a key factor to measure its comprehensive strength. As the leading force of domestic high-quality development enterprises, the company should take advantage of the trend, actively help realize the "double carbon" goal, taking green low-carbon business to an opportunity to practice the important strategy of low-carbon construction. The low-carbon thinking application is the key to urban development. The company has rich experience in urban comprehensive development. The difficulty of implementing green and low-carbon technology lies in the determination of the general direction of technology application and the integration of new strategies. This study mainly puts forward suggestions for updating the management model from two aspects: the development direction of enterprises and the integration of new green and low-carbon construction contents.

Business management, adhere to the "department independence, business integration" work philosophy, all departments work basically independent, and development positioning and work objectives coordination and unity, all departments in order to perform their functions on the basis of the need to further strengthen the department cooperation, to achieve the circulation of resources and information sharing, joint efforts into the development direction of the enterprise. On this basis, focusing on the development activities of green and low-carbon urban areas, the research established a three-level management system of planning guidance, collaborative support and business management, enriched the green and low-carbon technology control links, strengthened the coordination and cooperation between departments, and guaranteed the implementation process by improving the management system.

## References


Huan L, Hongbing D (2021) Study on the spatiotemporal differences in the coordinated development of population urbanization and land urbanization in the Yangtze River Economic Belt, China (2016-5):160–166

Changhui W, Lei W (2016) Analyze the management mode of the project company during the construction period of PPP projects. China Water Transport Construction Industry Association enginnering construction commitee 2016 anunual confernce proceedings (1):173–178

Fuya C, Jiaxin M (2021) Research on performance management of PPP project in China. Res Econ Manag (1)

Zifang J (2018) Application of internal control in project management companies. Account Learn (11):2

Mei Z, Xianjin H (2021) Research on urban carbon emission accounting and influencing factors in China (2019-9):13–19

Taohong W, Zhe S (2022) Low-carbon transition and green innovation: evidence from pilot cities in China. Sustainability 14

Hua Z (2016) Research on collaborative management of financial management and cost control in real estate enterprises. Account Learn (1):2
Danyang Y (2014) Research on improvement strategies for construction cost control and management of building engineering. Real Estate Guide 000(31):399 (2014)
Yuan F (2019) The impact of new PPP management policy on construction enterprises and their coping strategies. Urban Archit 16(14):169–170
Wu Q (2015) CSCEC equation: exploring the PPP Model of Tanggu Bay and starting a new city. Architecture (15):2
Zhaoyu Z (2021) On construction management and quality control of building electrical engineering. Clausius Sci Press (5)
Jiawei W (2020) Research on urban comprehensive development capacity of construction enterprises under PPP mode. Constr Econ 41(6):3
Chenglong L (2019) Research on the performance management of infrastructure projects based on PPP model. Fort Today (20):2
Zhiguo L (2014) LY Real estate company M project cost management research[D]. Jinan:Shandong University (2014)

# Chapter 11
# Current Situation, Dilemma and Path Selection of Construction Waste Treatment in China

**Yanyan Wang, Lijun Qi, and Han Cai**

**Abstract** With the acceleration of the urban process, in order to solve the problem of producing a large number of construction waste, resource utilization is the most reasonable solution. This paper first introduces the current situation of urban construction waste treatment, construction waste recycling industry contains multiple governance subjects and new models of governance, in spite of this, construction waste recycling in the development of the country is still faced with many difficulties. In view of the difficulties faced by urban construction waste treatment, this paper puts forward corresponding countermeasures, especially the commercial operation mode of the main body of treatment, to build the industrial chain of construction waste recycling, and promote the development of construction waste recycling industry.

**Keywords** Construction waste · Recycling · Governance subject · Operation mode

## 11.1 Introduction

In recent years, China's government has paid more and more attention to the concept of green development, and is guiding various industries to transform and upgrade, changing the past high energy consumption and high emission production methods, and achieving the goal of "carbon peaking" and "carbon neutral". In the face of the problem of construction waste management, China adopts a governance system combining unified supervision by the central government and hierarchical management by regional government departments, with the central government often formulating guiding management policies and governments at all levels formulating subdivision decrees according to the guiding documents and the actual situation in their

Y. Wang · L. Qi (✉)
Shandong Jianzhu University, Jinan, China
e-mail: 643049282@qq.com

H. Cai
Nanjing University of Science and Technology, Nanjing, China



P. Mendonça et al. (eds.), *Proceedings of 2023 International Conference on Green Building*, Environmental Science and Engineering,
https://doi.org/10.1007/978-3-031-43478-5_12

own regions to carry out construction waste management. Not only that, the main body of construction waste management needs to change to a model of continuous collaboration among multiple subjects (Song 2019). The collaborative governance mechanism of multiple subjects involving government departments, social organizations and enterprises, and the public can become a new path to solve the dilemma of construction waste management (Li 2018). This paper attempts to analyze the current situation and dilemma of urban construction waste governance in the context of collaboration among multiple subjects, and propose corresponding optimization paths to solve the dilemma of multiple governance of construction waste from the demand of national modern governance.

## 11.2 Current Situation of Urban Construction Waste Treatment

### *11.2.1 The Government Actively Introduces Regulation and Encouragement Policies*

In addition to strong regulatory policies, the government has also formulated tax policies and subsidy policies to encourage construction waste treatment enterprises and resource-based enterprises to actively participate in the construction waste treatment industry and promote the development of recycled products. For example, construction waste recycling products are given priority in engineering projects involving government funds; VAT reduction and income tax exemption policies are implemented for construction waste resourcing enterprises; the proportion of recycled aggregates in renewable resource products sold exceeds 90% to receive certain VAT reduction and exemption benefits.

### *11.2.2 Increasingly Prominent Role of the Market*

With the social and economic development, the past environmental governance system led by the government's administration has gradually failed to meet the demand for collaboration among multiple subjects. The main body that generates construction waste is the market, and only when the market is involved in the governance work can the problem be solved at root (Lu 2020). Relevant enterprises in the construction waste management industry include architectural design companies, construction enterprises, building demolition enterprises, waste transportation enterprises, landfill enterprises, as well as construction recycled material sales enterprises and resource recovery enterprises. With the guidance of various incentive policies and the rise of PPP (Public–Private-Partnership: to open up infrastructure construction and public service projects to social capital.) financing methods, more and more enterprises are entering the construction waste management industry. However, the utilization and industrialization of recycled products in China are still in the development stage.

### 11.2.3 Social Organizations Provide More Professional Services

Construction waste management is a process that combines hard institutional constraints and soft cultural constraints, which requires strict supervision by policies and social organizations to solve the problem of insufficient citizen participation consciously. The introduction of social organizations as the "third hand" of public governance can, to a certain extent, make up for the shortcomings of the government and the market, change the government-led monolithic service model, and correct market failures (Wang 2019). In recent years, in addition to social organizations formed spontaneously in each region, associations and committees in the industry have also actively promoted the positive interaction between government governance and social regulation. For example, the Urban Environment Association and the Construction Waste Management and Resourcefulness Working Committee hold annual national construction waste management exchange conferences, with participants including the Ministry of Housing and Construction, the Development and Reform Commission, the Ministry of Ecology and Environment, and local government departments, as well as relevant universities and experts in the industry.

### 11.2.4 Research Institutes and Universities Are Committed to Studying New Governance Models

As the main participants of construction waste management, research institutes and universities have the same rights and interests as the public to enjoy a good environment, and at the same time, they can use their knowledge power to improve the environment. Many universities in China cooperate with government departments, social organizations and enterprises to jointly promote construction waste management, and at the same time actively work with them to carry out international forums and devote themselves to improving construction waste management.

### *11.2.5 The Media and the Public Gradually Participate in Construction Waste Management*

Publicizing construction waste management activities and exchange conferences, popularizing sustainable concepts to the public and exposing illegal dumping of construction waste are inseparable from media channels. As a communication bridge between the public, government, social organizations and enterprises, the media can quickly generate public opinion influence. The public is one of the beneficiary subjects of construction waste management and has the right to enjoy a good living environment. Therefore, both the media and the public are the supervisors of construction waste management. In places where government supervision is not timely, the public can report and complain about enterprises and projects that handle construction waste illegally through the media.

## 11.3 Difficulties Faced by Urban Construction Waste Treatment

### *11.3.1 The On-site Classification System of Construction Waste Needs to Be Strengthened*

For construction waste, on-site sorting followed by transportation, treatment and resource recycling is the most efficient method. There is no mandatory on-site sorting system for construction waste, and on-site sorting is costly and less profitable, so the social responsibility of construction enterprises alone is not enough to motivate them to take the responsibility of sorting; and the huge "secondary sorting" workload of unclassified waste for disposal and resource recovery enterprises will greatly increase their labor costs (Tang and Liu 2016). In China, the lack of awareness of construction waste sorting among market players, the lack of quality assurance of sorting, and the inadequacy of sorting safeguards have affected the overall efficiency of construction waste management.

### *11.3.2 Imperfect Information Communication and Supervision Mechanism Information*

Information communication and supervision are mainly through the information management platform, administrative approval platform, location monitoring platform, process control platform, site control platform, big data platform and APP of construction waste disposal established by each region, and the establishment of the platform helps construction waste information sharing and collaborative supervision. However, in actual operation, due to the different degrees of development among cities and the different degrees of importance attached to environmental management by local departments, the degree of information disclosure through the platform varies greatly, the content of information disclosure is incomplete and the information is not updated in a timely manner, and the public participation in the platform is less applied and the role of the platform is not fully played.

### 11.3.3 Lack of Guiding Evaluation Standards for Recycled Products

The industry evaluation standards currently developed in China focus on low value-added construction waste recycling products, including recycled aggregates, recycled concrete, etc. The utilization path of low value-added products is restricted and faces the problem of narrow sales path. The evaluation standards for high value-added construction waste recycling products are still missing, including organic products, light aggregate products, 3D printing materials and other products. The lack of industry and national evaluation standards for high value-added recycled products will result in low acceptance, which is not conducive to relevant enterprises to attract investment. Failure to establish a sound evaluation standard for recycled products circulating in the market is likely to cause undue competition and exacerbate the level of distrust in recycled products.

### 11.3.4 It Is Difficult to Balance the Interests of Governance Subjects, and Difficult to Cooperate

The different interests of various governance subjects and the inconsistent and difficult to balance final goals make it difficult for the collaboration of construction waste governance subjects. Stakeholders have different attitudes and initiatives towards construction waste management based on their own interests. Social organizations are not independent enough, most of them are official or semi-official, and they need to rely on the government's assistance in the process of obtaining information and resources. When they collaborate with the government, the market and the public, it is difficult to get reasonable decentralization from the government, which makes their ability to allocate social resources poor. The grassroots social organizations responsible for construction waste management have fewer members and have to undertake various tasks such as supervision, propaganda and coordination, which greatly affect the efficiency of work. At the same time, the government has the right to choose social services, and the public can only passively accept public services provided by social organizations; and the public welfare activities formed by the public spontaneously lack proper promotion channels, resulting in the inefficiency of cooperative governance of construction waste management.

## 11.4 The Path of Urban Construction Waste Treatment

### *11.4.1 Policies and Regulations on On-site Classification of Construction Waste Shall Be Issued*

Government departments should stipulate the rules of construction waste sorting and require relevant enterprises to strictly implement on-site sorting and disposal of construction waste, which can refer to the relevant standards of European countries for detailed sorting of construction waste; secondly, a standardized sorting and disposal process should be established, and relevant enterprises should comply with it and gradually adopt mechanized sorting for operation; government departments can grasp the on-site sorting Then, the government should provide subsidies and incentives to enterprises that implement construction waste sorting according to the regulations; it can also build a construction waste sorting and point reward system based on blockchain architecture, etc., taking reference from the form of domestic waste treatment; finally, the planning of construction waste sorting centers should be integrated with the construction of recycling centers to meet the requirements of each function.

### *11.4.2 Improve Information Communication and Supervision Mechanisms*

Accelerating the process of establishing platforms such as information management platform, administrative approval platform and APP for construction waste disposal is conducive to the rapid development of construction waste resource-based industry. For example, the management of construction waste on-site classification and transportation should be strengthened, and a series of monitoring systems for construction waste demolition, transportation and treatment should be improved through the comprehensive use of Beidou positioning technology and intelligent monitoring, so as to synchronize information on the location of construction waste transportation vehicles and the type and quantity of waste in real time. In addition, the government should establish a high-level platform for the transformation and promotion of results, and the relevant departments should promote the establishment of waste recycling technology transfer demonstration institutions for research institutes, build a service bridge for long-term dialogue between research institutes and resource-oriented enterprises, help research institutes to conduct reasonable and professional pricing evaluation of relevant results, and help the reasonable transformation of construction waste recycling products. On the other hand, big data platform and APP etc. can constantly push the importance of construction waste reuse and the latest development of the industry, and promote and educate the public on construction waste related knowledge.

### *11.4.3 Establish Evaluation Standards for Effective Recycled Products*

Construction waste recycled product evaluation standards should be jointly prepared by industry standards committees, universities and scientific research institutes, etc., by the Ministry of Housing and Urban–Rural Development Institute of Standards and Quotations, professors from universities and other industry experts composed of the expert committee to apply for investigation, multi-faceted multi-angle review of the industry standards for recycled products in the practical application of whether it is feasible. Evaluation criteria should specify different kinds of construction waste recycled products terminology, definitions, specifications, classification and product marking raw materials, technical requirements, test methods, inspection rules, signs, packaging, storage and transportation, including the main raw materials to which the standard applies. It will not only help to have specifications for construction waste recycled products, but also break the barrier of poor marketing brought by the previous lack of specifications and promote the utilization of recycled products.

### *11.4.4 Construct the Business Operation Mode of the Governing Body*

The government formulates laws and regulations for construction waste disposal and constructs a reasonable supporting policy framework, which can make construction waste resourceization a key content and make clear provisions for the control standards and specific implementation operations in all aspects of waste management, especially to propose evaluation standards for recycled product quality and on-site classification system for construction waste. The construction of an industrial chain from construction waste to recycled products, which includes transportation enterprises, resourceization enterprises, scientific research institutions, universities and social organizations, and finally get recycled products from construction waste. PPP projects can be included in the industrial chain to inject capital resources for industrial development. At the same time, the media and the public are gradually involved in construction waste governance, and a communication and supervision platform is built to commit to information transparency and sharing in the construction waste resourceization industry. Clarify the role of the node in the development of

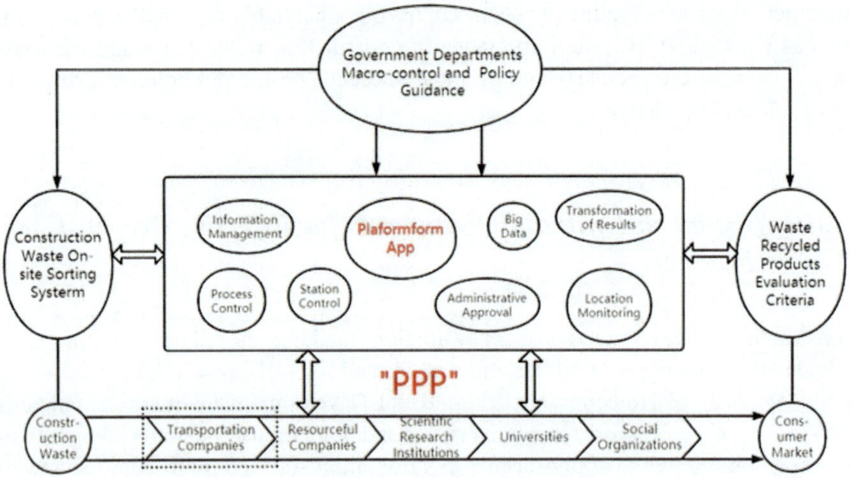


**Fig. 11.1** Commercial operation mode of the main body of governance

the resource-based industry and build a business operation model for the governance body (Fig. 11.1).

## 11.5 Conclusion

The development of construction waste recycling industry is related to the development of politics and economy. Waste reproduction is a systematic project involving all links of waste generation, transportation and reuse, which requires the full cooperation and participation of all relevant departments. In the development of construction waste recycling industry, the government should issue relevant laws and regulations or introduce PPP projects to give full financial support to scientific research institutions, recycling enterprises and transportation enterprises, and all enterprises should actively participate in the development of construction waste recycling industry. At the same time, all relevant subjects should actively promote the use of recycled products and improve the public's recognition by using the communication and supervision platform. The commercial operation mode of the main body of governance should be established, and all links should coordinate and effectively interact, so as to form a closed loop for the development of the construction waste recycling industry and truly realize the recycling.

## References

Li T (2018) Collaborative governance: a review of domestic research and extraterritorial progress. Soc Res (03):131–143

Lu Q (2020) Research on the connotation and realization path of regional environmental collaborative governance. Theor Perspect (02):59–64

Song H (2019) The construction of public safety governance model from the perspective of collaborative governance. Changbai J (04):65–72

Tang G, Liu X (2016) Exploring the collaboration mode of multiple subjects of rural science and technology services—Jiangsu rural science and technology service supermarket as an example. China Sci Technol Forum (08):137–142

Wang Y (2019) Research on sustainable development of social organizations under the perspective of urban governance. Urban Dev Res 26(05):81–85

# Chapter 12
# Investigation, Analysis, and Application of the Greening Landscape of Qushuiting Street in Jinan Old City District

**Yiwei Xiao**

**Abstract** Street greenery is not only the backbone of urban greenery, but also an important part of the street landscape, playing a crucial role in shaping the city's image. As one of the most representative historical and cultural districts in Jinan, Qushuiting Street is an outstanding display of the profound heritage and unique features of Jinan. Due to the special geographical location and the overall water layout, Qushuiting Street, forms a natural and smooth street space and unique landscape greenery, which brings a good space experience for citizens and tourists. Through data collection and field investigation, this article conducts a research and analysis of the current status of street greenery in Qu Shui Ting Street, a historical and cultural old street in Jinan City, Shandong Province, from the aspects of development and evolution, spatial pattern, and historical value. At the same time, according to the actual data statistics and its usage, it evaluates and summarizes the landscape greenery of Qushuiting Street based on the current situation.

**Keywords** Street Greenery · Qushuiting Street · Investigation

## 12.1 Overview of the Historical and Cultural Block of Qu Shui Ting Street

Jinan, the capital of Shandong Province and the birthplace of "Longshan Culture", has more than 2700 years of history (Shurun 2019). As a famous historical and cultural city in China, Jinan has a rich cultural heritage. After a long historical development, Jinan has formed many ancient streets and alleys at different times, which not only vividly tell the past, but also deeply influence the present and the future.

Qu Shui Ting Street in Jinan is one of the famous historical and cultural characteristic old streets. The street is located in the old city of Jinan, with a length of 530 m,

Y. Xiao (✉)
Civil and Architecture Institute, University of Jinan, 336 Nanxinzhuang West Road, Jinan City, Shandong, China



P. Mendonça et al. (eds.), *Proceedings of 2023 International Conference on Green Building*, Environmental Science and Engineering,
https://doi.org/10.1007/978-3-031-43478-5_13

north of Daming Lake, south of Xigeng Road, east of the North Gate of Dewang Palace, and west of Jinan Prefecture Study and Wen Temple. Today, in the depths of this bustling city, this small street still reproduces the ancient style and charm of the past, interprets the market situation of the past, and tells the story of an old city.

### 12.1.1 History of Qushuiting Street

The history of Qushuiting Street can be traced back to the Northern Wei Dynasty, with relevant records in the "Commentary on the Water Classic" by Li Daoyuan. After the Song Dynasty, due to its proximity to the county and prefecture government, this area became the core of the old city of Jinan. During the Ming and Qing dynasties, it was under the jurisdiction of the Dewang Mansion. Two small bridges, Qifeng Bridge and Baihua Bridge, were built at the intersection of Qushui and Yudai Rivers, and an open pavilion was built on the river bank, which was later called Qushuiting. In the late Qing and the early Republic of China, with the continuous development of the old city, the prosperous scene of Qushuiting Street became a permanent memory of old Jinan after the commercial center of Jinan shifted. Today's Qushuiting Street may no longer have the scene of "clear springs flowing on the stones," but the green weeping willows on both sides of the Qushui River and the various lifestyles of residents drinking tea and chatting on the bank are all telling us that this is the old Jinan that people remember (Chengyang et al. 2019).

### 12.1.2 Unique Street and Alley Structure of Qushuiting Street

The street is the basic framework of urban space and the most important form of expression for the spatial form of historical and cultural districts. The Qushuiting Street not only serves as a transportation hub but also the main public space for daily activities of urban residents. As the Pearl Spring and Wangfu Pond converge to form the Quanshui River, the main street of Qushuiting Street forms a relatively regular space with a smooth direction. The buildings on both sides of the river are built along the river and are distributed according to the flow of the river. The streets and alleys are mostly in a T-shape with a high-to-width ratio of about 1:1. Some of the internal streets have buildings arranged in a staggered front-to-back layout, creating a unique sawtooth-shaped space (Xingming 2011).

### *12.1.3 Architecture and Landscape Features of Qushuiting Street*

The street's layout, which is in sync with the natural flow of spring water, features fluid curves that create a harmonious and flexible atmosphere. The wider west bank serves as the main road to solve the traffic of the streets, while the narrower east bank provides a pathway for local residents to travel. Most buildings on the street are quadrangle courtyards, with shops on both sides of the east and west banks (Na 2019). The main aterials of the building are grey brick and white stone, the roof form is sloping roof of green tile ecorated with grey tile, has a large warping. The riverbanks are adorned with two rows of weeping willows that sway in the wind, while the clear water allows the water grass to sway with the water flow. The green landscape of Qushuiting Street serves practical functions within the space created by the streets and buildings and enables people to better experience the historical atmosphere of the city.

## 12.2 Analysis of the Current Situation of Greening on Qushuiting Street

The greenery design of Qushuiting Street is well done with beautiful landscapes and reasonable plant arrangements. On both sides of the street, various trees, flowers and plants are dotted, creating a pleasant green landscape. The plant arrangements not only consider the beautification of the landscape, but also take into account the needs of the urban environment and residents' lives.

### *12.2.1 History of Qushuiting Street*

The plant species found in the surveyed area were counted and classified. The trees planted along the street include weeping willows, elms, dryland willows, honeysuckle, wintergreen, and parthenocissus tricuspidata, using a total of 5 families, 5 genera, and 6 plant species, including 5 woody plants and 1 herbaceous plant. Among the woody plants, there are 3 types of trees, and shrubs and vines are 1 type each. The frequency analysis of the plant application shows that weeping willow has the highest usage frequency as a roadside tree, with more than 40 occurrences, mainly distributed along the road and riverbank. Other plants are individually planted at the street corners or in front of the shops on the east bank. See Table 12.1 for survey results.

**Table 12.1** Plant species' statistics and status investigation and analysis

| Tree species | Willow | Elm | Dry willow | Honeysuckle | Euonymus ilex | Virginia creeper |
|---|---|---|---|---|---|---|
| Quantity | 42 | 1 | 1 | 1 | 1 | 2 |
| Height | 10-15m | 3m | 9m | 4m | 2m | 3m |
| Growth condition | Excellent | Good | Poor | Excellent | Excellent | Good |
| Type | Deciduous' tree | Deciduous' tree | Deciduous' tree | Herbaceous plant | Bush | Liane |
| Position | Both sides of the road | Street corner | Roadside | The shop front | Street corner | The shop front |

### *12.2.2 Plant Configuration and Greening Landscape Analysis*

According to different properties, functions, and scales of the road, the greening section is divided into one plate and two strips, two plates and three strips, three plates and four strips, four plates and five strips, and other forms (Chengshan 2005). Qushuiting Street adopts a one-plate-two-strip form, with one side adjacent to the water and the other side connected to the buildings. The roadside trees are arranged on the main road along the riverbank, and the main configuration form is the tree pit type, mainly planting weeping willows.

The overall greening form and level are simple and have seasonal changes. In spring, Qushuiting Street is full of greenery, with the weeping willow branches swaying in the breeze. In winter, only the bare branches of the willow trees can be seen drooping down. Although it is difficult to see the charming scene of "water and willows everywhere" in the traditional sense, one can still feel the original flavor of old Jinan in the swaying willow branches.

### *12.2.3 Greening Landscape Effect*

After investigation, the greenery planting area of Qushuiting Street covers the entire street from north to south, first distributed evenly on both sides of the road, forming a green "roof" covering the street. Then, heading south, there are single-row weeping willows planted along the Qushui River, combined with two stone bridges connecting the east and west banks. The thick and curved willows always lightly brush the water surface with their flexible branches, and strolling on the stone-paved riverside path, one can feel as if they are in a garden of Jiangnan, experiencing the picturesque scenery of the south.

## 12.3 Evaluation of Greenery Utilization in Qushuiting Street

The street greenery of Qushuiting Street plays a very important role in landscape effect, crowd demand, and urban green organization, making the street space more representative and historically significant.After questionnaire survey on different people in Qushuiting Street, the evaluation of street greening in Qushuiting Street is summarized as follows.

### *12.3.1 Meeting the Requirements of the Street's Nature and Functionality*

Historic and cultural neighborhoods are important carriers of cities' historical changes and cultural memories, carrying part of cities' residential and commercial functions. Qushuiting Street is one of the most distinctive commercial streets in Jinan, where visitors can not only experience unique folk culture, but also taste Jinan's delicious food and snacks. The gathering of residents and tourists at specific times makes the old street crowded, and traffic pressure increases synchronously. However, the trees on both sides of the road, while ensuring the shading function, do not affect the narrow street width, but instead define the street space of Qushuiting Street and ensure the openness of the street.

### *12.3.2 Effectively Improve the Level of Street Greenery*

In the large-scale "old city renovation" movement in Jinan, the street greenery of Qushuiting Street has always been protected as a unique feature and has been relatively intact (Yi 2012). In recent years, with the government's investment and public support, the greening renovation of old city streets has achieved significant results, effectively changing the outdated greenery appearance of the old city area and greatly improving the landscape quality of the central urban streets.

### *12.3.3 Creating a Unique Urban Image*

Located in the central urban landscape belt of Jinan, Qushuiting Street's long history and unique style are excellent tourism resources that are rare in modern cities. In the city's development process, it has retained the unique regional environment, cultural characteristics, and architectural style of the city, shaping Jinan's ancient city style that has been around for a thousand years. Whenever people think of the willow catkins blowing in the breeze and the willow branches hanging on the bank, they will think of the winding ancient streets and alleys of Jinan, full of greenery and old-world charm.

### *12.3.4 Good Landscape and Ecological Benefits of Street Greenery*

Street greenery focuses on ecological benefits and follows the principle of sustainable development. The street greenery of Qushuiting Street connects the "points" and "faces" in the urban green space in a strip shape, forming the "vein" of the urban green space system (Siyong et al. 2013), which is a vital component of the urban ecological system. The ecological benefits produced by the street greenery make the street landscape cleaner, healthier, and more natural, and have a positive impact on people's senses, psychology, and emotions, satisfying people's inherent affinity and dependence on nature.

## 12.4 Conclusion

As one of the few remaining historical neighborhoods in Jinan, Qushuiting Street embodies the value of old streets in terms of history, culture, architecture, and street layout, to some extent reflecting the traditional urban landscape and cultural heritage of Jinan.

According to the methods of data collection and field investigation, the article studied and analyzed the current situation of the landscape greenery of Qushuiting Street, as well as through interviewed and surveyed on the feelings of local residents, tourists, merchant and office workers on the greening landscape. It is concluded that the use of green landscape in Qushuiting street is favourable, which brings good experience to people. In short, the study on the street greening of Qushuiting Street in Jinan City not only plays an important role in cultural protection and tourism development in Jinan, but also has important implications for social and cultural development, urban planning, and the protection and inheritance of traditional culture.

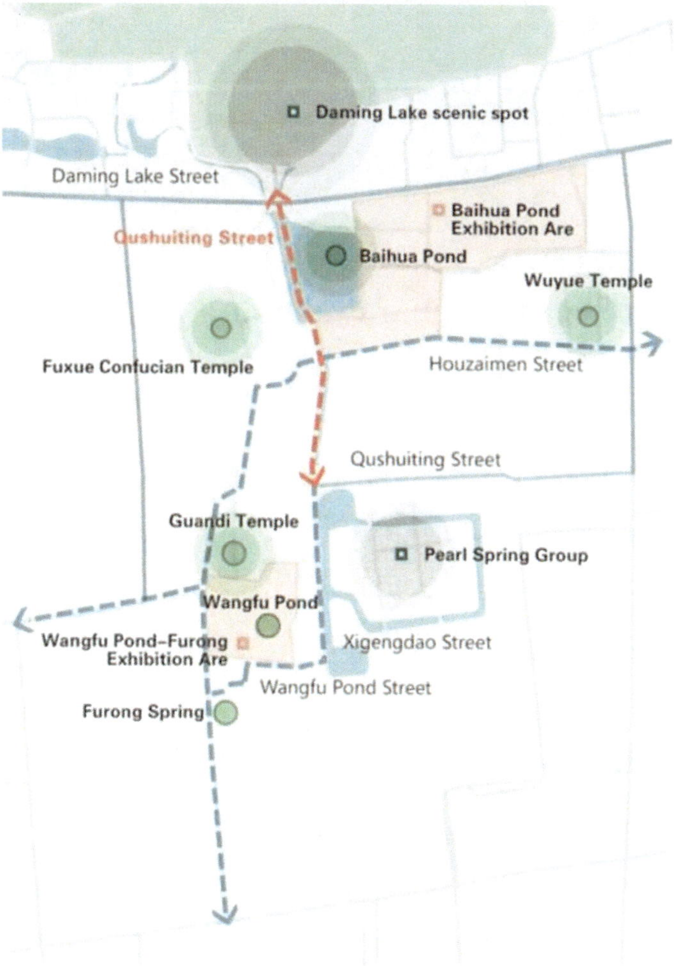


**Fig. 12.1** Comprehensive cultural landscape resources of the historic district

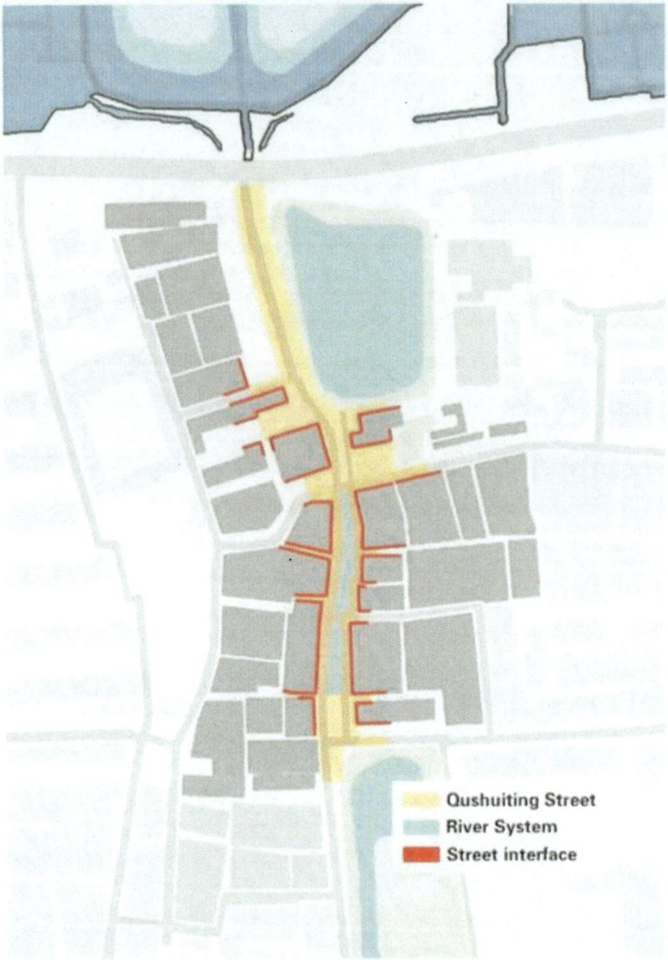


**Fig. 12.2** The road direction and building interface of Qushuiting Street

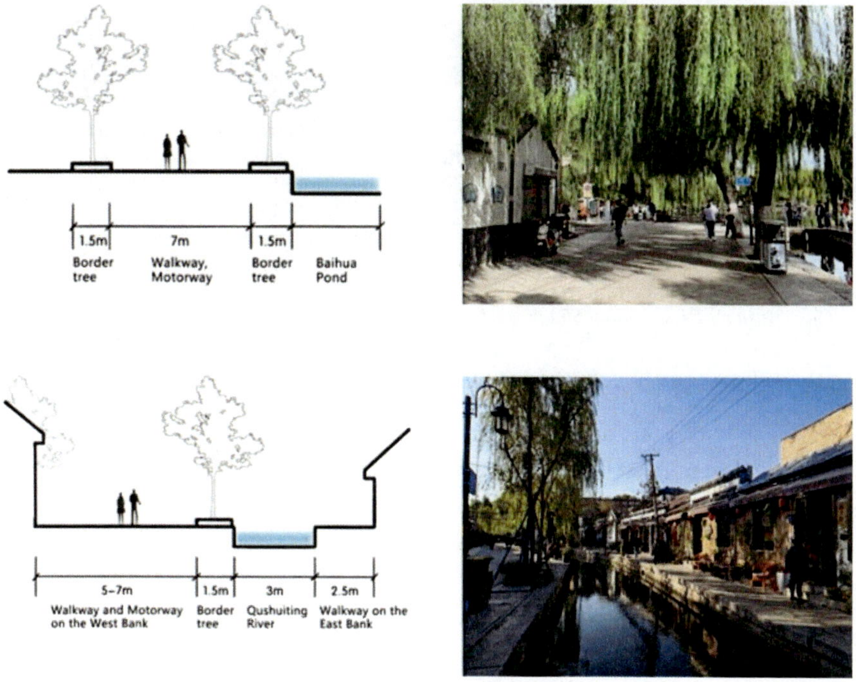


**Fig. 12.3** Two different forms of green sections in Qushuiting Street and their corresponding status photos

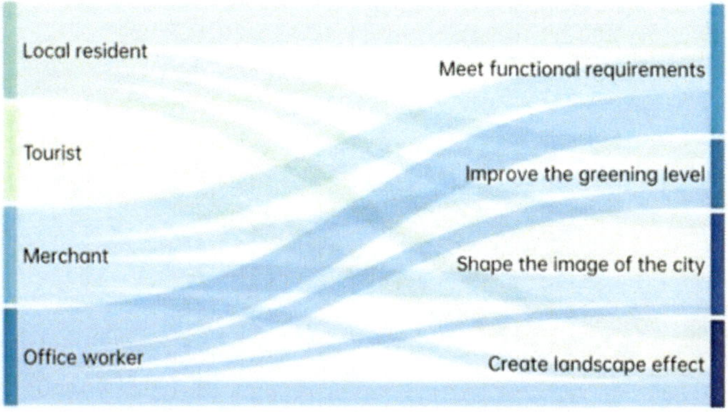


**Fig. 12.4** Evaluation on the use of street greening in Qushuiting Street by different people

## References

Chengshan L (2005) Investigation and analysis of street greening in Songjiang central city of Shanghai. *Master's Thesis, Zhejiang University*

Chengyang S, Shengge W, Tianbin S (2019) Spatial form of ancient street in Qushuiting street, Jinan. Build Mater Decor 06:91–92

Na S (2019) Historical preservation and commercial development of urban old blocks. Qushuiting Street in Jinan. Farm Staff (03):233

Shurun W (2019) The protection and significance of regional characteristics of Jinan traditional Dwellings in Urban renewal. IOP Conf Series: Mater Sci Eng 592:012108

Siyong Y, Yuanyuan F, Ruidong S, Yujing W (2013) Investigation and analysis of street green space in downtown street in Handan city. J Green Sci Techn 09:105–106

Xingming Z (2011) Vigorous renaissance of historic precincts: a case study of Furong historic precinct in Jinan. Mod Urban Res 01:44–48

Yi H (2012) Research on the protection and development of historical and cultural districts in Jinan City. *Master's Dissertation, Shandong University*

# Chapter 13
# Analysis of the Wind Environment in the Building Process of Marine City

**Longlong Zhang, Jingwen Yuan, and Chulsoo Kim**

**Abstract** The buildings of marine cities have to consider very factors in the construction process, among which the wind environment is crucial. Extracting the architectural system of buildings adapted to the regional environment and interpreting and modifying it scientifically by using modern materials and technologies has an important inspirational value and significance for the development of modern green building industry. By studying the wind environment system in the built environment of marine cities, this paper is a rethinking of the modern green building panoply in marine cities, providing ideas for the moderate development of green building systems in coastal cities today, and realizing the synergistic development of traditional green buildings and modern green technologies.

**Keywords** Wind · City environment · Marine city · Building

## 13.1 Introduction

The healthy and sustainable development of ocean cities plays a role of agglomeration, leading, attraction and radiation to the political, economic, cultural and social development in the region, thus driving the hinterland economy and forming a development pattern of land and sea (Saunders 2003). Marine cities are urban areas located along the coast or on islands, and are exposed to unique environmental conditions. Wind is one of the most important of these conditions, as it can have a significant impact on the construction process. Construction in marine cities is further complicated by factors such as limited space, challenging terrain, and exposure to natural disasters such as hurricanes and tsunamis. Therefore, understanding and predicting the wind environment is crucial for ensuring the safety and sustainability of marine city construction projects.

L. Zhang · J. Yuan · C. Kim (✉)
Department of Marine Design Convergence Engineering, Pukyong National University, 48513 Busan, Republic of Korea
e-mail: bosen@pukyong.ac.kr



P. Mendonça et al. (eds.), *Proceedings of 2023 International Conference on Green Building*, Environmental Science and Engineering,
https://doi.org/10.1007/978-3-031-43478-5_14

The main objective of this study is to improve our understanding and prediction of the construction process in marine cities through a comprehensive analysis of the wind environment in marine cities. Specific objectives include:

To assess the impact of the wind environment on building projects in marine cities, including the effects on the stability of building structures, energy consumption and ecology.

To provide targeted and optimised design and response measures to reduce the adverse effects of the wind environment on building projects by comparing and analysing the wind environment in different maritime cities.

Identify gaps in current research and directions for future research to provide guidance for subsequent studies.

The sustainable urban architectural environment is the environmental capital of the city, and the healthy and sustainable development of the architectural environment is the core of the ecological carrying capacity of the city. The living spatial layout of the city is called the spatial form of the city, and this spatial form has a certain impact on the ecological balance and the quality of resources of the city. Therefore, the study of sustainable and optimal design of the architectural environment in ocean cities is conducive to the healthy and environmentally friendly development of urban health through the social, physical and psychological aspects of the architectural environment (Zhang et al. 2023).

Through this study, we will be able to achieve the following results:

Understand the characteristics and variation patterns of the wind environment in marine cities: through field observations and numerical simulations, we will obtain a detailed understanding of the wind environment in marine cities, including the variation patterns of parameters such as wind speed, wind direction and wind frequency.

Assessing the impact of the wind environment on construction projects in marine cities: By assessing the impact of the wind environment on construction projects, we will be able to propose targeted design optimisations and countermeasures to reduce the adverse impact of the wind environment on construction projects.

## 13.2 Experimental Analysis

### *13.2.1 Case Location and Data Collection*

Suzhou is located in the middle of the Yangtze River Delta, with a history of over a thousand years of development, and is close to the first-tier developed cities of Shanghai and Zhejiang. It has a low topography and a rich water environment, with 36.6% of the city's land area covered by water alone, making it a famous water town in the south of the Yangtze River. Suzhou has a subtropical monsoonal maritime climate, with mild weather throughout the year, four distinct seasons, abundant rainfall and long frost-free periods. The average monthly temperature for the last five years on

**Fig. 13.1** Suzhou area map

the Suzhou Meteorological Network shows that the lowest temperatures of the year occur in January and December, with an average temperature of 5.4°. The highest temperatures are found in July and August, with an average temperature of 27.8° and above, as shown in Fig. 13.1.

### *13.2.2 Methodology*

The following methods will be used in this experiment:

Field observations: Representative locations in the maritime city will be selected for observations of the wind environment, including parameters such as wind speed, wind direction and wind frequency. Factors such as urban topography, building height and building layout will also be measured and recorded.

Numerical simulation: The wind environment of the maritime city is simulated and predicted using meteorological models and numerical simulation techniques. Mathematical models are used to simulate changes in the wind environment under different conditions and their possible effects on building projects.

The experimental steps are as follows:

Selection of the study site: A representative maritime city is selected as the study object and observation sites in the city are selected.

Data collection: Field observations are carried out at the selected observation sites to collect data on the wind environment, including parameters such as wind speed, wind direction and wind frequency. Factors such as urban topography, building height and building layout are also measured and recorded.

Data processing and analysis: The field observation data are processed and analysed, and numerical simulation techniques are used to create mathematical models to simulate the changes in the wind environment under different conditions and to analyse their possible effects on the building project.

Presentation of results and conclusions: Based on the experimental results, the characteristics and changing patterns of the wind environment in marine cities will be demonstrated, their impact on the construction process will be assessed, and optimised design and countermeasures will be proposed.

### 13.2.3 Selection of Relevant Parameters

The thesis will apply CFD simulation techniques to the analysis of the area. In order to make the best use of the computer and software to simulate the wind environment, it is necessary to make the relevant constraints on the working conditions used in this thesis. It involves the selection of wind speed, the setting of boundary conditions such as inlets, and the determination of relevant elements such as geometric modelling and meshing, and the setting of the computational domain (Kim et al. 2021).

(1) Surface roughness, wind direction, wind speed and related parameters setting

In the bottom layer of the atmosphere, due to the frictional effect of the surface, the wind speed at the surface will increase or decrease with the change of height above the ground. It is generally believed that the wind speed is not affected by the surface until 300–500 m above the ground, and the wind speed varies along the height from the ground to the interval not affected by the surface. Its formula is shown in Eq. (1).

$$\frac{U_z}{U_0} = (\frac{Z}{Z_0})^a \tag{13.1}$$

In the formula: $U_z$ represents the wind speed at a height of $Z$ metres from the ground in m/s; $Z_0$ represents the height from the ground of the measuring point of the wind speed measured by the weather station, generally taken as 10 m; $U_0$ is the wind speed measured by the weather station and a is a constant reflecting the roughness of the ground (Kim et al. 2021). In our specification, the terrain is divided into four different types, A, B, C and D. The corresponding values are shown in Table 13.1.

As the thesis simulates mostly urban areas with dense building clusters, a ground roughness a of 0.22 is taken and in high density urban centres, a ground roughness a of 0.3 is taken for the simulation.

The selection of parameters associated with the wind simulation includes the setting of wind direction and speed, temperature, etc. The climate in a city area is hot in summer and cold in winter, with four distinct seasons. Therefore, in the planning and design, ventilation in summer and shelter from the wind in winter must be taken into account by analysing the meteorological parameters of a city. Parameters such

**Table 13.1** Surface roughness index classification table

| Category | Areas covered | Roughness index |
|---|---|---|
| A | Offshore seas, islands, coasts, lakeshores and deserts | 0.12 |
| B | Suburbs of large cities with open fields, countryside, jungles, hills and small to medium-sized towns with sparse housing | 0.16 |
| C | Urban areas of cities with dense clusters of buildings | 0.22 |
| D | Urban areas of large cities with a high concentration of buildings and a large number of high-rise buildings | 0.3 |

as the dominant wind direction and the average wind speed in summer and winter are determined (Yuk et al. 2020).

### *13.2.4 Analysis of Results*

On the basis of the experiments, the following conclusions can be drawn:

The wind environment in marine cities has significant spatial and temporal variability and requires classified research and the development of countermeasures for different geographical and climatic conditions.

The impact of the wind environment on marine city construction projects is multifaceted and requires comprehensive optimisation measures to be considered during the design, construction and operation phases.

There are still gaps and shortcomings in the current research, and further in-depth research is needed on the formation mechanisms, impact modes and optimisation of countermeasures for the wind environment in marine cities.

If the same coastal city is located in a different region, the dominant preference wind direction is also different, and the traditional concept of sitting north facing south is not always a favorable orientation. For example, in Guangzhou, the favorable orientation is 15° to the east of south and 5° to the west of south; in Xiamen, the dominant wind in summer is southeast, and the best orientation for buildings is 5–10° to the east of south, while the favorable orientation of Suzhou is 15° to the east of south. The wind shadow area and vortex length formed by the wind surface of the building are related to the building's external dimensions. So the relationship between the dominant wind direction and the building orientation should be handled well. Research shows that when the orientation of the building takes the incidence angle of the dominant wind direction in summer between 30° and 60°, which is conducive to the smooth natural ventilation. From the vertical layout of the building, when the ranks and rows of buildings behind are high, the building should avoid facing the dominant wind in winter, otherwise the vortex zone generated is too large, which will have a negative impact on the natural ventilation of the buildings in the back row. If it is impossible to avoid the vortex area, public toilets or cleaning buildings should be avoided in the vortex area, otherwise the dirty air cannot be discharged, leading to the deterioration of environmental health. Considering the above factors, for a group of buildings arranged in a regular row, their orientation should be at a certain angle to the prevailing wind direction in order to let the wind blow to each building (Yuk et al. 2020).

## 13.3 Conclusions

In the construction of marine city buildings, many locations has gradually embarked on the right track in the path of sustainable development, various effective policies and regulations have been released one after another to guide, and a perfect theoretical system has been gradually built up. This paper combines wind environments to conduct an in-depth study of green building optimisation, and analyses the role of wind environments on building design optimisation. A general framework of the building water environment system is proposed based on artificial intelligence technology, and the wind environment is simulated to provide ideas for building design (Hassan 2018).

Meanwhile, with the continuous construction of buildings in coastal cities, it is necessary to evaluate the wind environment in the overall planning of building groups and the design of building units, and to optimize the design of buildings in terms of wind environment problems by means of wind tunnel simulation tests or computer numerical simulations. The design should consider the relationship between the wind environment system, the building system and the landscape system, and need to deal with the relationship between the subsystems of wind environment system of the marine city building, and then make scientific process selection and combination, and build the sustainable wind environment system of the green building. Reasonable architectural design forms create a comfortable natural ventilation environment and reasonable water system, while comfortable natural ventilation forms an ecological building energy-saving environment, and it can achieve a harmonious relationship between the form of architectural design and sustainable wind system, which is conducive to the development of the overall system of marine city, and ultimately promote the healthy development of the city (Lu 2015).

Although significant progress has been made in understanding and addressing the wind environment in marine city building processes, there are still several research gaps that need to be filled. Firstly, further research is needed to refine our understanding of wind erosion processes and their impact on building envelopes. Additionally, more work is needed to optimize wind energy systems for marine cities, ensuring their environmental sustainability. Furthermore, research is needed to determine best practices for integrating wind-resistant buildings into urban environments, in order to achieve sustainable urban development. Finally, research is also needed to address the economic viability of these measures, ensuring their widescale implementation in marine cities.

**Acknowledgements** This work was supported by a grant from Brain Korea 21 Program for Leading Universities and Students (BK21 FOUR) MADEC Marine Designeering Education Research Group.

## References

Hassan A (2018) The environmental geomorphological assessment of the urban expansion in Al-Khiran Marine City, Kuwait[J]. J Soc Sci Kuwait Univ 46(1):31–54

Kim J, Kang B, Kwon Y et al (2021) A review on the building wind impact through on-site monitoring in Haeundae Marine City: 2021 12th typhoon OMAIS case study[J]. J Ocean Eng Technol 35(6):414–425

Lu SM (2015) Marine cities: A vision for a future China[J]. Coast Cities Their Sustain Futur: 15–24

Saunders CA (2003) Ocean City-Longport replacement bridge requires precast concrete durability for Harsh marine conditions[J]

Yuk JH, Joh M, Huh T (2020) Simulation of storm wave run-up in the Busan Marine City, South Korea: A case study during Typhoon Chaba[J]. J Coast Res 95(SI): 1377–1382

Zhang L, Yuan J, Kim CS (2023) Application of energy-saving building's designing methods in marine cities[J]. Energy Rep 9:98–110

# Chapter 14
# Efficiency in the Preparation of Life Cycle Assessment

**Sina Hage, Sebastian Hollermann, Juliane Stelljes, Hermann Huber, and Timo Pakarinen**

**Abstract** The implementation of Life Cycle Assessments (LCA) in the planning process seems to be a challenge for most planners and is often not integrated into the planning process due to lack of time, interest or knowledge. The Building Information Modelling (BIM) method offers a promising basis for exchanging data more efficiently with LCA tools. Analyzing the current state of automation of this work process, the authors present a study about the preparation of a LCA with the open-source tool "eLCA" of the German federal institute for research on Building (BBSR) in terms of time efficiency and accuracy of two different workflows. The two workflows are the "Quantity take-off" method and the BIM based method "IFC import". A total of 66 participants from two different universities (Germany and Austria) attended the study. To measure the time efficiency of the specific workflow, participants are guided through a designed process using the process management tool "Camunda". The results of the study show that the method of "IFC import" allows on average a faster preparation of the LCA, but deviates more in the accuracy of the results. Moreover, this paper highlights the difficulties in preparing an LCA for first-time users.

**Keywords** Life cycle assessment · Process modelling · Building information modelling

S. Hage (✉) · S. Hollermann · J. Stelljes
Jade University of Applied Science, 26121 Oldenburg, Germany
e-mail: sina.hage@jade-hs.de

H. Huber
Salzburg University of Applied Science, 5431 Kuchl, Austria

T. Pakarinen
Karelia University of Applied Science, 80200 Joensuu, Finland



P. Mendonça et al. (eds.), *Proceedings of 2023 International Conference on Green Building*, Environmental Science and Engineering,
https://doi.org/10.1007/978-3-031-43478-5_15

## 14.1 Introduction

Our current emission and resource problem causes an ecological, sociological and economic crises. Climate change and the accompanying environmental disasters, in addition to sociological disasters, like poverty, migration, resource wars etc. are current and future challenges. Due to the high output of embodied energy and the high consumption of raw materials in the construction industry, architecture has a significant impact on the development of these crises. Currently, the construction sector is responsible for the largest consumption of resources in Germany, accounting for 22% of total resource consumption. 40% of all greenhouse gases in Germany are produced by the construction industry (Ramseier 2020).

In order to mitigate these threats, the creation of facts about the impact of building design, the interpretation of its facts and their interrelationships are of particular importance. Life cycle assessments (LCA) acts as a tool for creating details about the environmental impact of planned buildings or already existing ones and can thus support in decision-making by creating more sustainable buildings. Due to the time and cost involved in preparing an LCA, it is infrequent for a planned building, that a LCA will be prepared (Díaz and Antön 2014; Lambertz et al. 2019; Zimmermann et al. 2021). Certainly, a lack of know-how in the corresponding planning also contributes to the hesitation in dealing with LCA tools (Horn et al. 2020). This leads to the fact that LCAs are predominantly one-time prepared when it is required i.e., for a building certification, such as DGNB or BNB (Gantner et al. 2018; Hollberg et al. 2020; Xue et al. 2021). Nevertheless, to achieve the government's climate change goals, awareness of environmental impact factors is to be widely created. Whether the LCA is used as a tool in the design process as a basis for decision-making or as confirmation at the end of the planning process, the work steps of preparing a LCA must be simplified and accelerated, to shorten the process and to make the LCA analysis accessible to a wider audience without much LCA knowledge.

The currently widely used "Quantity take-off" method in Germany for preparing an LCA compiles the relevant building information in a corresponding Excel table (Santos et al. 2019). This requires several steps and is therefore time-consuming. The method Building Information Modelling (BIM) and the within developed standard data exchange format Industry Foundation Classes (IFC) of "buildingSMART" (Startseite | buildingSMART Deutschland, 2023), offer a promising basis for exchanging data more efficiently with LCA tools and thus accelerating as well as simplifying the work process of an LCA (Hollberg et al. 2020; Chong et al. 2017; Soust-Verdaguer et al. 2017; Opoku et al. 2021). This raises the research question of whether the workflow of the "IFC Import" allows a shorter and simpler creation of the LCA, than the "quantity take-off" method or whether further developments are needed to overcome the time-consuming creation of an LCA. Therefore, the following study examines these two workflows for creating an LCA in terms of its time efficiency and manageability for first-time users.

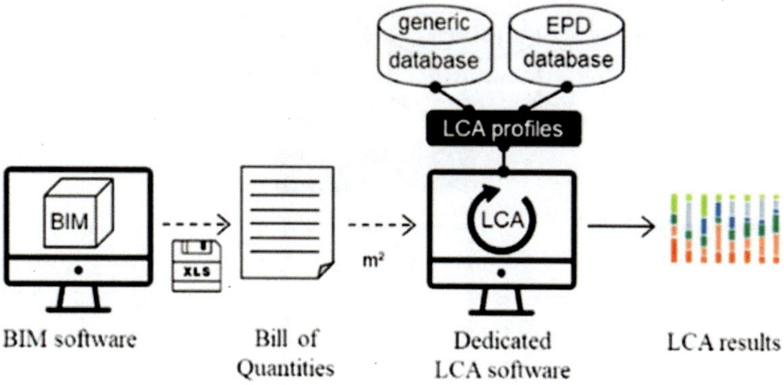


**Fig. 14.1** Quantity take-off "workflow scheme (Wastiels and Decuypere 2019)

## 14.2 LCA Workflow

In order to create an LCA, the masses of the planned materials of a building and other technical information, such as energy consumption of the facility equipment are necessary. There are several ways to transfer this information into LCA software. In this study, only two workflows are considered and compared for their time efficiency and manageability. The most common one in current design practice is the "Quantity-take-off" workflow and the new promising semi-automated is the "IFC-import" workflow. Both workflows are explained in the next paragraphs.

### *14.2.1 "Quantity Take-Off" Workflow*

The most common used workflow in current building design practice is the "Quantity take-off" workflow (Zimmermann et al. 2021; Wastiels and Decuypere 2019). In which the information will be based on a bill of quantities (BoQ) export refer to the following graphic (Horn et al. 2020), usually by using an Excel sheet, from a BIM model and which then have to be entered manually into a LCA software (Fig. 14.1). For this process, only the masses of the building components are provided. If the BIM model is to be supplemented with the results of the LCA, these must be entered into the BIM model manually.

### *14.2.2 IFC Import Workflow*

Another, promising in consideration of time efficiency, workflow is to create an LCA by importing the information of a BIM model with an IFC (Industry Foundation

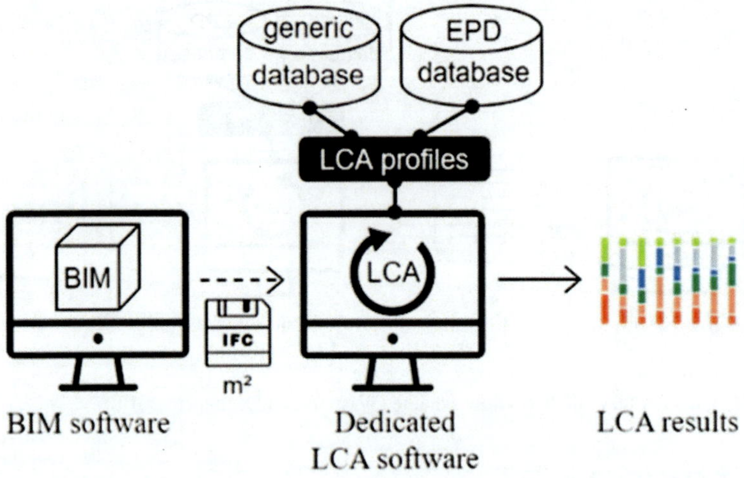


**Fig. 14.2** IFC import "workflow scheme (Wastiels and Decuypere 2019)

Class) file to the LCA software (Gantner et al. 2018; Su et al. 2020). The following graphic "Fig. 14.2" illustrate this workflow. Hollberg et al. (2020) study showed that "BIM can reduce the effort of calculating the embodied environmental impacts of buildings and therefore provides the potential to improve the environmental performance of buildings during the design stage" (Hollberg et al. 2020). As well as in the "Quantity-take-off" workflow the results of the LCA have to be reentered manually into the BIM model.

## 14.3 Methodology

The principal purpose of this paper is to measure and evaluate the processing time and challenges for first time users for the preparation of a LCA during the above mentioned two different workflows.

A very simple architectural design of a mobile classroom is taken as the basis of study. Plans with floor plan, section, elevations and details of the wall, roof and slab construction and the table of "Bill of Quantities" serve as the basis for the calculation. 66 students from a common course were divided into two groups for the respective workflow and assigned to work on the LCA. The evaluation of the participants' results was carried out using descriptive statistical method. The participants of the study were in their first or third bachelor's semester of civil engineering or construction informatics at the time of the workshop. The age of the 66 participants ranged from 18 to 29 years. The LCA study was conducted at two different university locations, in Salzburg, Austria and in Oldenburg, Germany.

At the beginning of each LCA Challenge, the basics of LCA were explained in a 60-min lecture in which the questions "What is an LCA?", "The purpose of an LCA?", "How to prepare an LCA?", "What information does an LCA provide?", "How are the results of an LCA to be interpreted?" were answered. Afterwards, the process flow was explained with the help of the tool Camunda, as well as the basics of the LCA software "eLCA". Then Login data for the Camunda interface was distributed to the participants by e-mail and with it, the participants were divided by random in two groups, "manually" and "BIM". The individual editing process could then be started and all further information was available via the Camunda platform in the workflow.

The open-source tool "bauteileditior" of the BBSR (federal institute for research on Building, Urban Affairs and Spatial Development) of Germany was used (BBSR 2022), which is according to EN15804 + A1 and linked to the German environmental impact database "Ökobaudat" (ÖKOBAUDAT 2021). Since February 2020 an IFC interface is available. This software was chosen because it is fully accessible to everyone. The BBSR thus offers a possibility for every planner in the building sector to create an LCA without making a large investment in the provision of the software and the training of the processors. This is a great advantage, especially for building planners with a low budget, in order to gain knowledge about the environmental impact of their building design.

To track the time and guide the participants through the specific workflow, the process management tool "Camunda" (The Universal Process Orchestrator 2022) was used. The following "Fig. 14.3" shows the created process chain. In the Camunda user interface the process chain was translated in the respective processing steps. The following Figures show the four workflow steps. In the first step questions were asked about the person and the existing experience with regard to the preparation of a life cycle assessment. The last question asks to which group the participant has been assigned. After selecting the appropriate group, one is directed to the specific process.

In each process, whether manual or BIM based, a link to the "bauteileditor" can be opened via the process. With given login data, users can log in to the respective profile. In the same step, the information on the building can be downloaded. The available information folder contains two plans, with floor plans, views and sections, as well as details on the component layers and information on the selected materials. In addition, an Excel table with the masses of the individual components is provided. Furthermore, a presentation is given that explains how to create a life cycle assessment

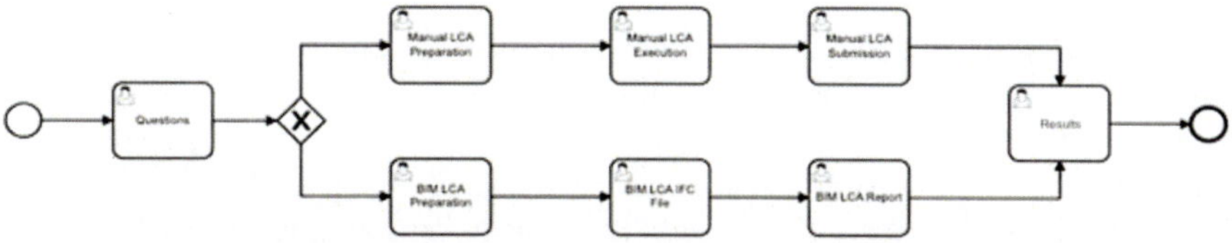


**Fig. 14.3** Camunda process model

in the “bauteileditor” and guides the user step by step through the process. In the process of BIM-supported creation of the LCA, there is also an IFC file for the building model in the download folder. After the information has been downloaded and the users have logged into the tool, the next step is to enter the time. For this, the user sets the current time and confirms it by clicking ok. The user is then invited to create the life cycle assessment. In the last step, the final report of the LCA can be uploaded to the Camunda Platform as a PDF file. Afterwards, the time is stopped by setting the time. In this way, the process ends.

To evaluate the data from the study, the data and information provided by the Camunda Process of each proband were transferred to an Excel spreadsheet and compared. For this purpose, the indicated time of the probands was presented next to the Global Warming Potential (GWP) value determined in the LCA. The GWP provides information on the amount of emitted greenhouse gases in the life cycle phases considered, such as production of the building materials and construction of the building (life cycle A), its operation (life cycle B) and the end-of-life scenario (life cycle C) (Global Warming Potential (GWP) 2023). In this study, the GWP was determined within the system boundary “cradle-to-gate” (life cyle consideration A to C). In addition to the GWP, other environmental impact indicators are determined in an LCA. However, in order to compare the accuracy of the respective LCAs results of the test participants, the GWP is used as the measured value.

## 14.4 Results

The groups of participants in each location were divided into two groups “Quantity take-off” or also called “manual workflow” and the “IFC Import” workflow or “BIM workflow”. Whereby everyone in the group had to create an LCA on their own. By monitoring the creation of the LCAs over time, it is expected to gain knowledge about which processing variant is more time-efficient. Furthermore, the supervision of the students during the processing allows insights into the manageability of the different processing procedures and reveals possible obstacles in the workflow. The results of the two locations, Salzburg (Austria) and Oldenburg (Germany) are presented below. In the first table of each location the number of participants in total and per group are demonstrated. It was noted that some participants have not uploaded a LCA report to the Camunda platform, but had a result in the LCA software. To show the handling with the process management tool, the tables show also four columns beside the total participants column with the following information:

*A = “with LCA report and result”:* the participant has uploaded a LCA report on Camunda platform and succeeded a result.

*B = “without LCA report, with result”:* the participant has not uploaded a LCA report on Camunda platform, but accomplished a result in the LCA software.

*C = “with LCA report, without result”:* the participant has uploaded a LCA report on Camunda platform, but without a LCA result neither in the LCA software.

*D = "without LCA report, without result":* the participant has not uploaded a LCA report neither accomplished a result in the LCA software.

### *14.4.1 Location Salzburg, Austria*

The study with Location Salzburg, Austria was conducted on 18.10.2022 at 3:00 pm with 29 participants in total. The following Table 14.1 shows the distribution of the number of participants in each group and how many of them reached a result, as explained before in Chap. 4.

The following Box plot diagram (Fig. 14.4. Result "LCA editing time"—Group Salzburg, Austria) shows the time-based comparison of the two considered workflows. It shows that the probands obtained a result on average 16 min faster with the "IFC Import"/BIM method. It also illustrates that, considering the standard deviation, the method "IFC Import"/BIM allows a faster processing of the LCA.

Considering the accuracy of the results on average, the participants of the manual method obtained a GWP (Global Warming Potential) by means of 11.24 kg $CO_2$eq/ $m^2$a in a time of 1:37 h. The standard deviation of the GWP result is 0.69. The participants of the BIM method have calculated a GWP of 13.26 kg $CO_2$eq/$m^2$a at a time of 1:21 h, here the standard deviation of the GWP result is 3.96. The following diagram (Fig. 14.5) compares the two workflows in terms of the accuracy of the results and demonstrates that the "Quantity-take-of" method offers a more accurate method for determining greenhouse gases due to the lower standard deviation of the results among the subjects.

**Table 14.1** LCA challenge participants location A

| Location: Salzburg, Austria | Participants | A | B | C | D |
|---|---|---|---|---|---|
| Participants "Manual" workflow | 17 | 5 | 5 | 0 | 7 |
| Participants "BIM" workflow | 12 | 7 | 1 | 4 | 0 |
| Total | 29 | 12 | 6 | 4 | 7 |

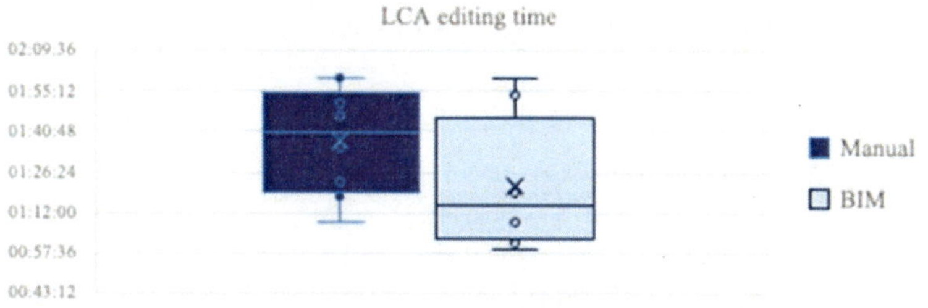


**Fig. 14.4** Result "LCA editing time"—Group Salzburg, Austria

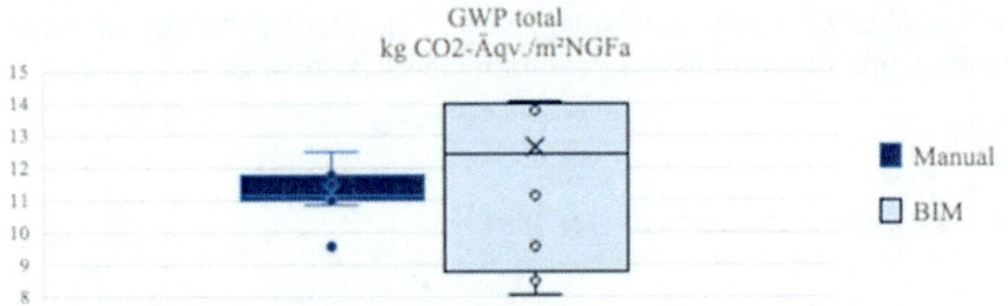


**Fig. 14.5** Result GWP total—Group Salzburg, Austria

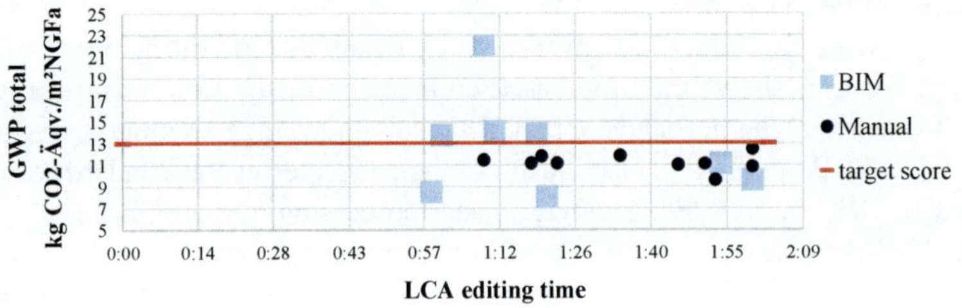


**Fig. 14.6** GWP/Time—Group Salzburg, Austria

The subsequent diagram (Fig. 14.6) shows the results of all test persons of both workflow groups in comparison with regard to their processing time and the GWP value determined. The red line shows the GWP value of the test building, which was calculated by sustainability experts. The following table shows the distribution of the number of participants in each group and how many of them reached a result, as explained before in Chap. 4.

## 14.4.2 Location Oldenburg, Germany

The study in Oldenburg, Germany was conducted on 15.12.2022 at 2:00 pm with 37 participants in total. The following Table 14.2 shows the distribution of the number of participants in each group and how many of them reached a result, as explained before in Chap. 4.

**Table 14.2** LCA challenge participants location B

| Location: A | Participants | A | B | C | D |
|---|---|---|---|---|---|
| Participants "Manual" Workflow | 18 | 13 | 4 | 0 | 1 |
| Participants "BIM" Workflow | 19 | 15 | 1 | 1 | 2 |
| Total | 37 | 28 | 5 | 1 | 3 |

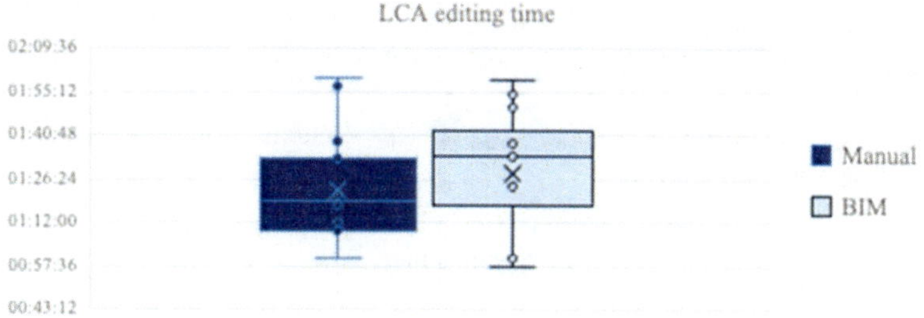


**Fig. 14.7** Result "LCA editing time"—Group Oldenburg, Germany

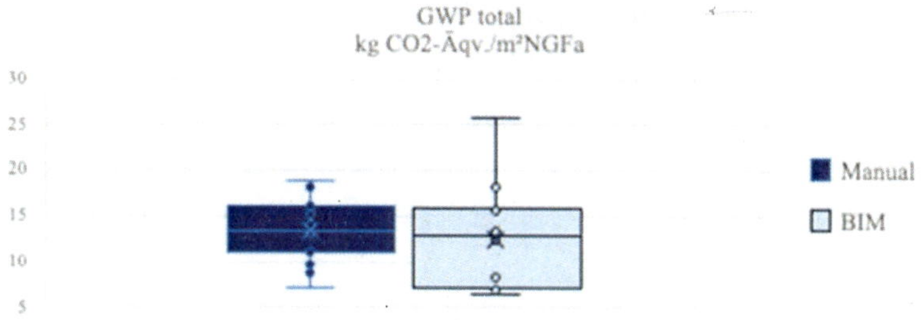


**Fig. 14.8** Result GWP total—Group Oldenburg, Germany

In contrast to the expected result that the "IFC Import"/BIM Method enables a faster work process, the test persons in Oldenburg need on average 5.5 min longer. This trend illustrates the following diagram Fig. 14.7.

On average, the participants of the manual method obtained a Global Warming Potential by means of 13.43 kg $CO_2eq/m^2a$ in a time of 1:22 h. The standard deviation of the GWP result is 3.32 kg $CO_2eq/m^2a$. The participants of the BIM method have calculated a GWP of 12.46 kg $CO_2eq/m^2a$ at a time of 1:28 h, here the standard deviation of the GWP result is 5.36 kg $CO_2eq/m^2a$ as it can be seen in the following "Fig. 14.8".

The following diagram (Fig. 14.9) shows the distribution of the individual results of the test persons with regard to time and the GWP achieved in the location Oldenburg, Germany. The red line shows the GWP value of the test building, which was calculated by sustainability experts.

### *14.4.3 Overall Consideration*

In the following section, the results of the two groups of participants from Germany and Austria are considered together. The box plot diagram below (Fig. 14.10) shows

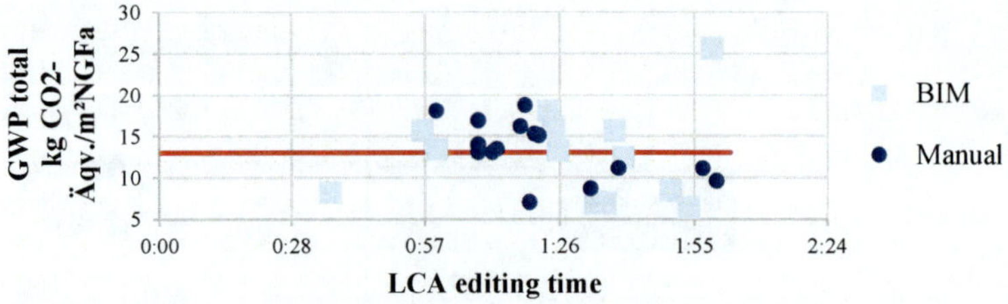


**Fig. 14.9** GWP/Time—Group Oldenburg, Germany

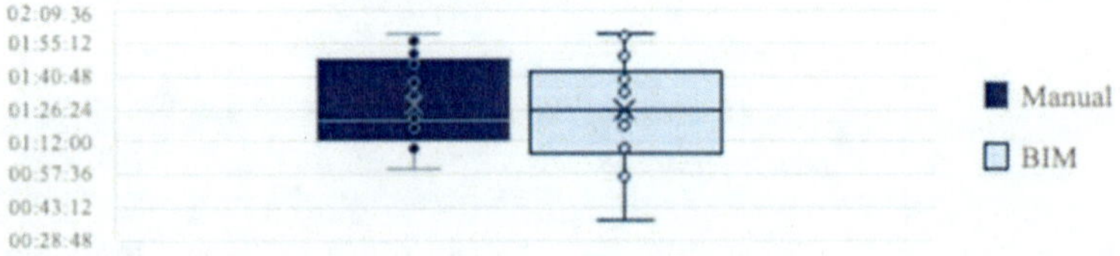


**Fig. 14.10** Result "LCA editing time"—total of all groups

the average time the participants spent on the respective workflow. The graph illustrates that the workflow with the BIM method is on average 3 min faster than the "manual" workflow.

In the subsequent comparison of the GWP results (Fig. 14.11) of all test persons in relation to the workflow, it becomes apparent that the standard deviation of the GWP results of the workflow with the BIM method is greater than that of the model variant.

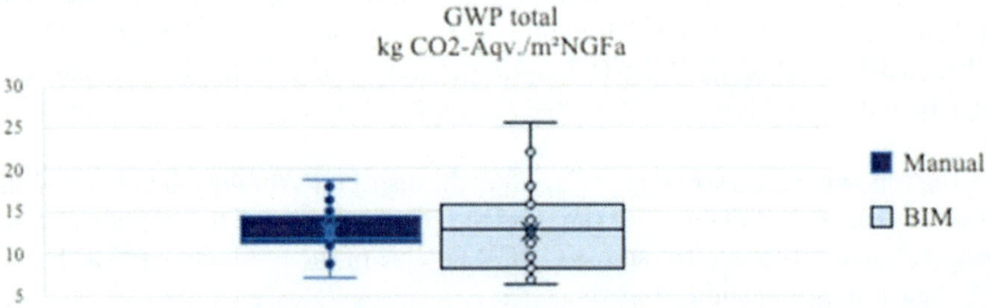


**Fig. 14.11** Result GWP total—total of all groups

## 14.5 Discussion

From the results of the workshop in Salzburg it can be seen that the test persons with the BIM method were on average 16 min faster than the test persons with the "manual" method. However, the workshop in Oldenburg does not show this trend; there, the test subjects were even 6 min slower on average. Overlapping the results of the Oldenburg and Salzburg group shows that the BIM method was carried out 3 min faster on average. Overall, this study shows that the integration of the BIM method can accelerate the work process of preparing an LCA. It is also observable that the variance of the GWP results are greater with the BIM/"IFC Import" method compared to the manual/"Quantity-take-off" method. During the practice it became apparent that the data of the model of the IFC file was not correctly transferred to the LCA software. This meant that the mass data had to be edited during the IFC import, which also cost time and leads to inaccuracy. From this, it can be assumed that an accurate IFC import can once again increase the efficiency of the LCA creation process by means of the BIM method. At the same time, it was noticed that the participants found it difficult to interpret the results and thus did not question the result further or check their input. An integration of benchmarks that automatically align to the entered building model could be helpful. It can also be assumed that by supposedly automating the process, the user hands over the responsibility of the result to the automation. However, also stopping the time could also have led to a subconscious time pressure among the participants, which triggered the ability to deliver a result as quickly as possible without checking it. Stopping the time in a hidden background and extending the general time of the workshops could exclude any subconscious of time pressure and so minimize the deviation of the GWP results. The different knowledge background of the participants, due to different study programs and age, may likewise have influenced the results. In particular, the process management via the process management tool "Camunda" required the participants to familiarize themselves with these tools. A possible difficulty in using this tool for some participants may have caused that only 40 (60.6%) out of a total of 66 participants fully completed the process on Camunda by uploading the final result and stopped their time. During the introductory lecture, it was observed that 97.8% of those involved had no knowledge about LCA or were familiar with the topic. The introduction was carried out in all locations with the same presentation slides. Nevertheless, it is possible that the results were influenced by a not exactly identical reproduction of the presentation contents. Furthermore, it should be noted that the example used is very small and the results could be different in terms of time savings with a larger building model and with test persons who are already familiar with the preparation of an LCA. An extension of the IFC file with the information of the individual component layers, referring to the IDs of the "Ökobaudat" would be a great relief. At the same time, the LCA software would have to be able to recognize these and create the components itself using the IDs.

All these conclusions provide a basis for further investigation of the efficiency of the specific workflows considered, on which further investigations can be carried out.

## 14.6 Conclusion

The study shows two possible workflows of information exchange with the LCA software "bauteileditor" with regard to their time efficiency and indicated challenges for first time users of the LCA software "eLCA" and their barriers during the workflow. In general, the study shows that the BIM method can save time in the work process of creating an LCA, but the accuracy of the transfer of information must be taken carefully into account. This provides a basis for further studies.

Expanding on this study, larger building models could be simulated to verify the time difference into the two workflows. In addition, this study can be repeated by using a different commercial LCA software, like "OneClickLCA" for the BIM method workflow.

## References

BBSR (2022). https://www.bbsr.bund.de/BBSR/EN/home/_node.html;jsessionid=5BE9AF0C4F900BE3262C0B9393241C81.live11313. Accessed 5 Jul 2022

Chong H-Y, Lee C-Y, Wang X (2017) A mixed review of the adoption of building information modelling (BIM) for sustainability. J Clean Prod 142:4114–4126. https://doi.org/10.1016/j.jclepro.2016.09.222

Díaz J, Antön LÁ (2014) Sustainable construction approach through integration of LCA and BIM tools, 283–290. https://doi.org/10.1061/9780784413616.036

Gantner J, Both P, Rexroth K, Ebertshäuser S, Horn R, Jorgji O, Schmid C, Fischer M (2018) Ökobilanz - Integration in den Entwurfsprozess: BIM-basierte entwurfsbegleitende Ökobilanz in frühen Phasen einer Integralen Gebäudeplanung. Bauphysik 40:286–297. https://doi.org/10.1002/bapi.201800016

Global Warming Potential (GWP) (2023). In: Statistisches Bundesamt. https://www.destatis.de/DE/Themen/Gesellschaft-Umwelt/Umwelt/Glossar/gwp.html. Accessed 17 Apr 2023

Hollberg A, Genova G, Habert G (2020) Evaluation of BIM-based LCA results for building design | Elsevier Enhanced Reader. https://doi.org/10.1016/j.autcon.2019.102972

Horn R, Ebertshäuser S, Di Bari R, Jorgji O, Traunspurger R, von Both P (2020) The BIM2LCA approach: an industry foundation classes (IFC)-based interface to integrate life cycle assessment in integral planning. Sustainability 12:6558. https://doi.org/10.3390/su12166558

Lambertz M, Theißen S, Höper J, Wimmer R, Meins-Becker A, Zibell M (2019) Ökobilanzierung und BIM im Nachhaltigen Bauen. Zukunft Bau, ein Forschungsprogramm des Bundesministeriums des Innern, für Bau und Heimat

ÖKOBAUDAT (2021). https://www.oekobaudat.de/en.html. Accessed 5 Jul 2022

Opoku D-GJ, Perera S, Osei-Kyei R, Rashidi M (2021) Digital twin application in the construction industry: a literature review. J Build Eng 40:102726. https://doi.org/10.1016/j.jobe.2021.102726

The Universal Process Orchestrator (2022). In: Camunda. https://camunda.com/. Accessed 15 Nov 2022

Ramseier L, Frischknecht R (2020) Umweltfußabdruck von Gebäuden in Deutschland, BBSR-Online-Publikation Nr. 17/2020, ISSN 1868-0097

Santos R, Costa AA, Silvestre JD, Pyl L (2019) Integration of LCA and LCC analysis within a BIM-based environment. Autom Constr 103:127–149. https://doi.org/10.1016/j.autcon.2019.02.011

Soust-Verdaguer B, Llatas C, García-Martínez A (2017) Critical review of bim-based LCA method to buildings. Energy and Buildings 136:110–120. https://doi.org/10.1016/j.enbuild.2016.12.009

Startseite | building SMART Deutschland. https://www.buildingsmart.de/. Accessed 17 Apr 2023

Su S, Wang Q, Han L, Hong J, Liu Z (2020) BIM-DLCA: An integrated dynamic environmental impact assessment model for buildings. Build Environ 183:107218. https://doi.org/10.1016/j.buildenv.2020.107218

Wastiels L, Decuypere R (2019) Identification and comparison of LCA-BIM integration strategies. IOP Conf Ser: Earth Environ Sci 323:012101. https://doi.org/10.1088/1755-1315/323/1/012101

Xue K, Hossain MU, Liu M, Ma M, Zhang Y, Hu M, Chen X, Cao G (2021) BIM integrated LCA for promoting circular economy towards sustainable construction: an analytical review. Sustainability 13:1310. https://doi.org/10.3390/su13031310

Zimmermann RK, Bruhn S, Birgisdóttir H (2021) BIM-based life cycle assessment of buildings—an investigation of industry practice and needs. Sustainability 13:5455. https://doi.org/10.3390/su13105455

# Chapter 15 Architecture and Sustainability—Recovering Green Areas in the Construction Process

**Joana Teles, Paulo Mendonça, and Ricardo Mateus**

**Abstract** This research explores the environmental impact of architecture and the need for sustainability in building construction. Currently, economic costs often overshadow long-term considerations. However, the concept of sustainability aims to change this mindset by emphasizing material choices, energy efficiency, and the preservation of green spaces and water management. The study focuses on restoring vegetation areas within building sites, seeking architectural solutions that offer benefits beyond mere decoration. The objective is to increase the density of developments while ensuring the preservation of green areas. Additionally, the research examines the potential impact on community habits and proposes home farming as a means to foster a closer connection with nature within urban buildings.

**Keywords** Sustainability · Home farming · Environment · Energy efficiency · Recycling

## 15.1 Introduction

Architecture has a great environmental impact because of the number of buildings that are constructed and the area they occupy. We live in a time where cities are growing with an increasing number of buildings and consequent reduction of vegetation. The current construction does not compensate in any way for the green areas occupied. The imbalance between built-up areas and green areas has negative consequences,

J. Teles (✉)
School of Architecture, Art and Design, University of Minho, Guimarães, Portugal
e-mail: joanasteles@gmail.com

P. Mendonça
Lab2PT, School of Architecture, Art and Design, University of Minho, Guimarães, Portugal

R. Mateus
Department of Civil Engineering, School of Engineering, University of Minho, Guimarães, Portugal



P. Mendonça et al. (eds.), *Proceedings of 2023 International Conference on Green Building*, Environmental Science and Engineering,
https://doi.org/10.1007/978-3-031-43478-5_16

resulting in decreased air quality, the occurrence of floods caused by soil sealing, and people's disconnection with the natural environment.

The aim of this document is to identify the adaptations needed to current architecture to make it more sustainable and to improve the quality of life of city dwellers, especially in the urban context. Referring to Fig. 15.1, It is a pertinent and necessary issue that the replacement of vegetation that is removed from the lot under construction, be done in the building to be constructed. This is a proposal to mitigate the negative impact of buildings on the urban context. The goal is to present architectural solutions to support and implement green areas in residential buildings, without taking too much space from the living area, ensuring that they have not only a decorative effect but also a functional one, thus transforming the primary function of shelter into a living space with many more valences. This research intends to quantify the potential for increased energy efficiency, resulting from the natural shading of the façades, the improvement of air quality (presence of greenery leads to the process of evapotranspiration) and the contribution to the quality of life of the inhabitants in general by analyzing social contacts and the concept of urban gardens for growing vegetables, herbs and fruits (cultivation). It is also intended to prove that the chosen solutions can be self-sufficient, not increasing the consumption of drinking water and contributing to the reduction of the negative effects of extreme weather phenomena. This research finally introduces a project example where the green area, equivalent to the area of its implementation, is recovered in the building, to study the results from the point of view of environment, sustainability, efficiency, resilience and habitability, comparing them with the current situation.

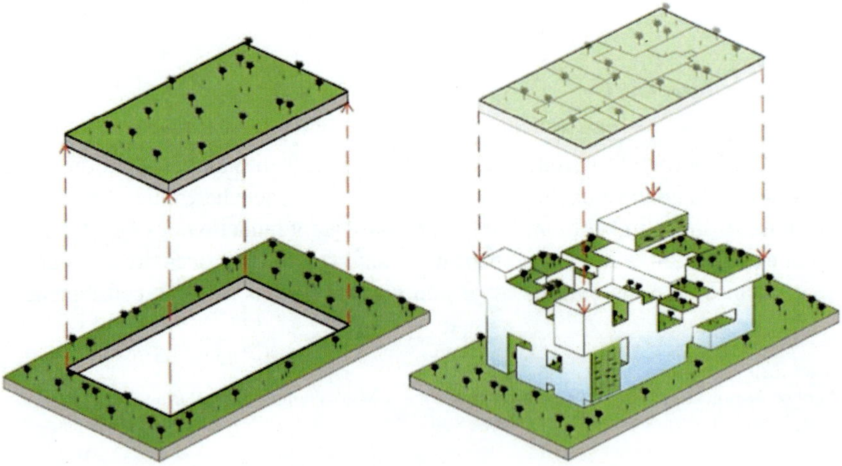

**Fig. 15.1** Replacing green area in an implementation building

## 15.2 Vegetation in Buildings and Environmental Responsibility

We are at a point in history where environmental and ecological awareness are issues of great importance in the most different areas, provided by events such as the climate crisis in which we find ourselves. The resolution of them in favor of more sustainable practices is one of the main focuses of the latest research in various fields (European Conference 1994). By modifying the water balance of the urban surface and the evapotranspiration process, we are contributing to the increased vulnerability of the population to the risks of floods, landslides, heat waves, among others. In this context, buildings can contribute to the mitigation of some of these problems, Fig. 15.2, promoting environmental integration in cities and better management of resources such as water and energy.

There are various ways to implement vegetation in the buildings such as Green roofs, High rise construction, in which one of the floors may be dedicated to the replacement of this green area occupied with vegetation, Vertical gardens and Domestic gardens, referred in Fig. 15.3.

### *15.2.1 Green Roofs*

The green roof is a solution that takes advantage of an area that is usually wasted and with no associated program, bringing cultivation and leisure activities to the residents of the building (The International Greenroof Greenwell Projects Database 2013). Green roofing consists of a covering, in the form of vegetation, over a built-up surface that can be developed according to flat or sloping surfaces, which can be accessible. They bring a new use to these wasted spaces, which can be used as leisure areas, parks for children, food cultivation from domestic gardens, areas for sports activities, among others. In addition to these benefits for the inhabitants, green roofs contribute to mitigate the decrease of green spaces in urban areas, to store collected rainwater and to delay its return to the sewage systems (ITECONS 2015). The application of vegetation on the roof also ends up reflecting most of the solar radiation, rather than absorbing it, avoiding heat gains in the structure, providing energy savings

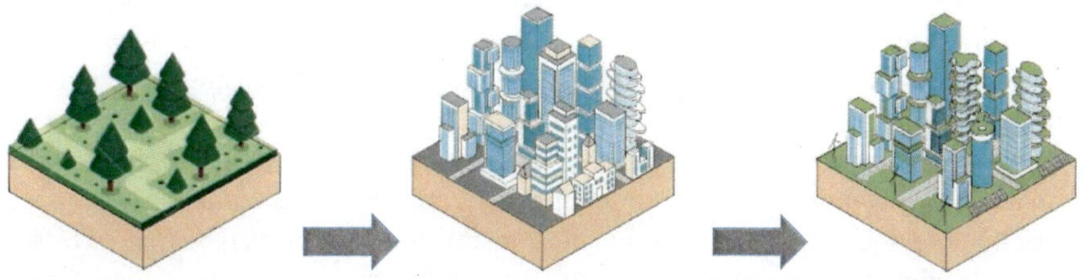

**Fig. 15.2** Green buildings = green city

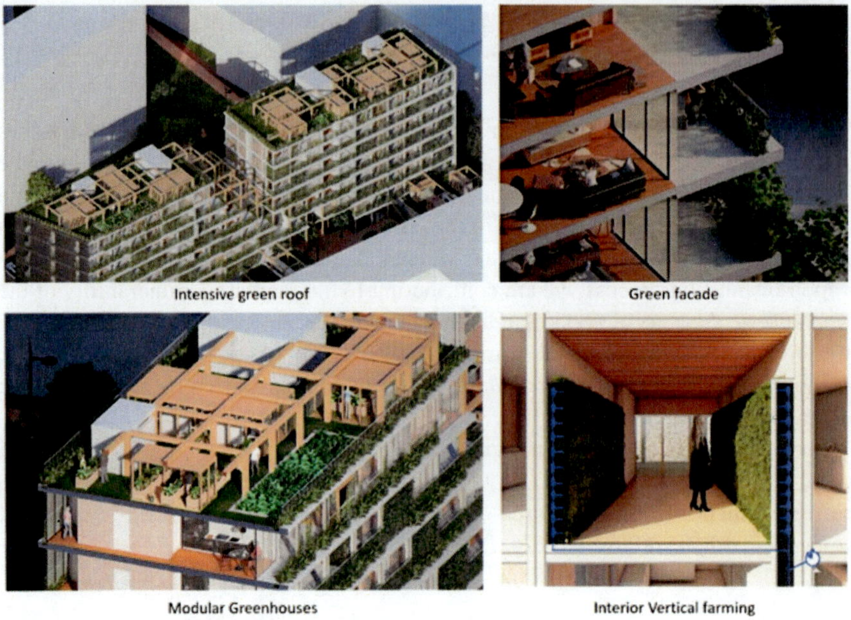


**Fig. 15.3** Integrated strategies in the demonstration building

for cooling the building. "In arid climates, the increased thermal inertia, due to the substrate, will increase comfort by reducing temperature fluctuations inside" (The International Greenroof Greenwell Projects Database 2013). The type of vegetation that will be applied to the roof will influence the thickness of the support slab. Green roofs can therefore be characterized into three types: Intensive, Simple Intensive and Extensive (RAPOSO, Fausto Miguel Ferreira 2013). Their types are specified on Table 15.1.

## *15.2.2 High-Rise Construction*

In urban regeneration, allowing the increase in height of the building with compensation of this new habitable area in green areas distributed over all floors of the building can be an interesting and economically viable solution. In this way the number of fractions is increased with a small reduction in the area of each, increasing the economic return of the building.

This solution has the advantage of increasing the housing supply without increasing the occupied area and, simultaneously, creating some compensatory green space that previously did not exist. In addition, this green space can be treated by the building's inhabitants themselves as a garden or cultivation area, constituting a catalyst for the human relations of its inhabitants. Obviously, green areas require water,

**Table 15.1** Green roof typologies

| Type of green roof | Vegetation variety | Depth of growth substrate (cm) |
|---|---|---|
| Extensive | Mosses—sedum | 4–8 |
| | Sedum—mosses—small herbaceous plants | 6–10 |
| | Sedum—mosses—small Gramineae herbaceous plants | 10–25 |
| Simple intensive | Gramineae herbaceous plants | 12 to >35 |
| | Wild bushes and shrubs—edible plants | 15 to >50 |
| | Medium wild bushes and shrubs—edible plants | 20 to >100 |
| Intensive | Grass | 15 to >35 |
| | Small bushes and shrubs—vegetable plants | 15 to >50 |
| | Big bushes and shrubs | 35 to >70 |
| | Large bushes and small trees | 60 to >125 |
| | Medium and big trees | 100 to >200 |

fertilizer, and, when artificial lighting is indispensable, electricity. Here appears a new area of focus for architecture where it is necessary to include infrastructures for rainwater collection, storage and reuse, and energy generation. These solutions also have application in new construction, where a significant area must be required to be returned to nature based on exactly the same principle—more height for green compensation.

### *15.2.3 Vertical Gardens*

Over time, new solutions for implementing green in buildings have emerged. As are the examples of green walls or living walls, also known as vertical gardens. Vertical garden was invented by Stanley Hart White who patented a green wall system in the late 1930s. A vertical garden is basically a self-sufficient living wall that is normally freestanding or attached to the outside/inside walls of a building (Blank 2008). This green wall can either function as a decorative feature, which may help with cooling the building, or as urban agriculture/urban gardening (Bennet 2013).

**Vertical Farming**. Vertical Farming is an economically cautious and efficient choice as it grows quickly, weighs little and occupies less land area. It allows faster and more controlled productions, irrespective of season, and doesn't require great food security (Simões 2020). There are three current vertical garden systems for controlled environment agriculture that do not require soil and only use nutrient water solutions to grow crops: Hydroponics, Aeroponics, Aquaponics.

Hydroponics is a system where the plant's roots grow in reservoirs and the nutrient solution is pumped around. Hydroponic farming could increase the yield per area of lettuce by around 11 times while requiring 13 times less water (Touliatos et al. 2016).

Aeroponics system is when the plant's roots grow free and the nutrient solution is sprayed directly onto them.

Aquaponics is a combination of hydroponics and aquaculture, where the plants use the fish waste as a fertilizer. The water is filtered when added again to the fish tank.

**Green Facade**. The vertical garden can also be used to create green facades (AMORIM, Ana Francisca Fernandes Ferreira 2015). By including a vertical structure design to cover specific places on the facade, the climbing plants can follow this structure according to growth and allow a vegetative look on the facade. Refering to the Fig. 15.4, the green facades can be made by using self-supporting plants, that do not need any support structure, like root climbers or adhesive-suckers, and by plants that require a supporting system like twining vines, leaf climbers or scrambling plants.

The green facade is also an excellent way to control the extreme temperature differences in Portugal, Fig. 15.5. Plants react to these changes by presenting little density in their foliage in winter, resulting in less heat absorption and shading effect, and denser in summer, resulting in more shading of the facades and cooling of the interior of the building.

**Home Farming**. Home farming is another topic to be addressed in the incorporation of green spaces in buildings. It is a kind of garden delimited by a specific area in the building. These places can be planned in the manner of the vertical gardens or they can adopt other growing systems using pots or even land. There are different methods that can be used to define a home garden. Users can have a space especially made for the activity by using the example of a module or a delimited area devoted to agriculture. This space can enjoy monitored irrigation systems, regulated energy, and the humidity and temperature necessary for food production (Stiles and Wootton-Beard 2017). Figure 15.6 demonstrates a project in which this system was used, where a number of modules were implemented with systems ranging from electricity generation, with solar equipment, waste management and recycling, and hydroponics, to serve the food cultivation activity of the building's inhabitants. Some of these modules were designed with the intention of being shared modules, promoting interaction and contact with the green areas and the people (PRECHT 2019).

Home gardens can also be designed in outdoor areas, such as balconies or roof tops, however the environmental issues of the space are linked to the climate and weather conditions of the outdoor space. This solution is perfect for short term and temporary cultivation. Home gardens require a specification in the construction of the building. Thus, the constructive detail must be considered for the implementation of these systems as well as the management of the energy and hydric components linked

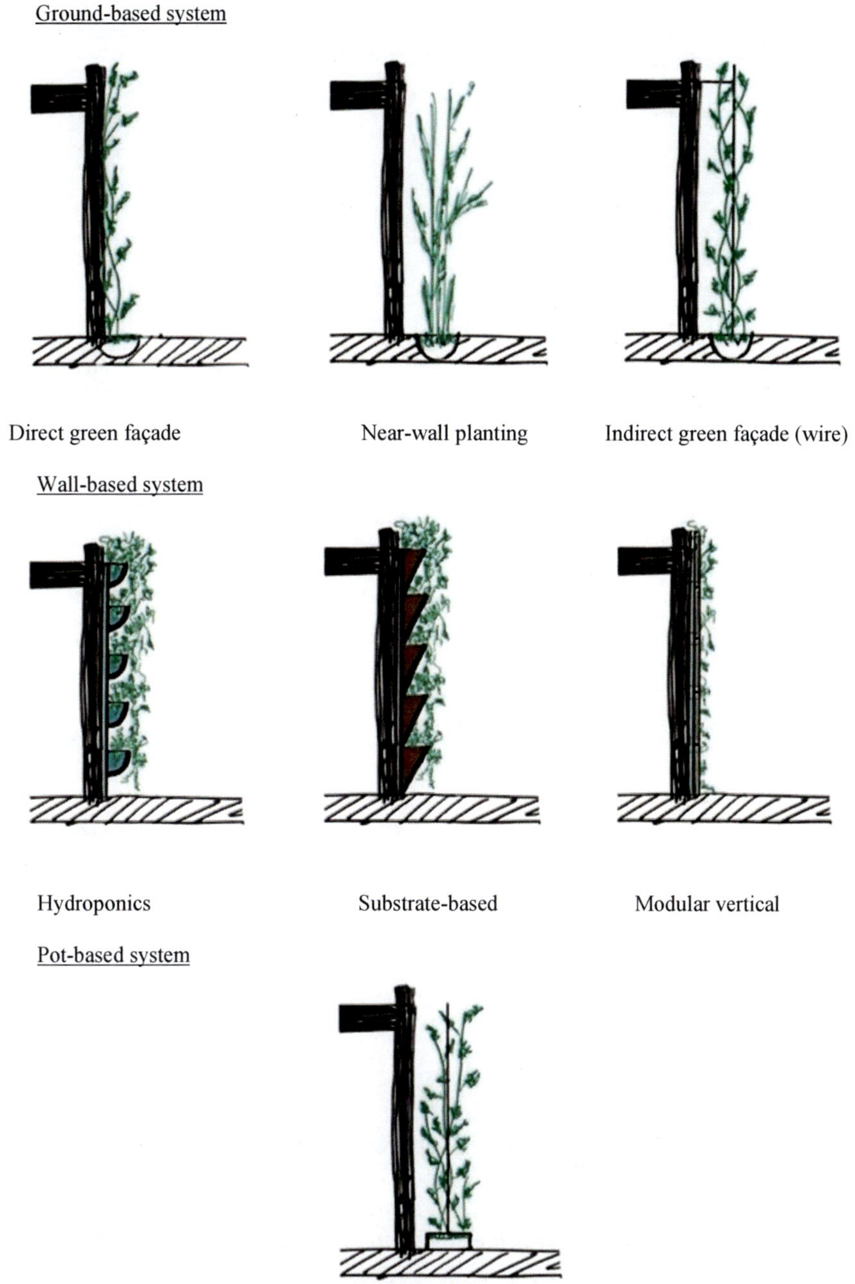


**Fig. 15.4** Classification of green facade systems

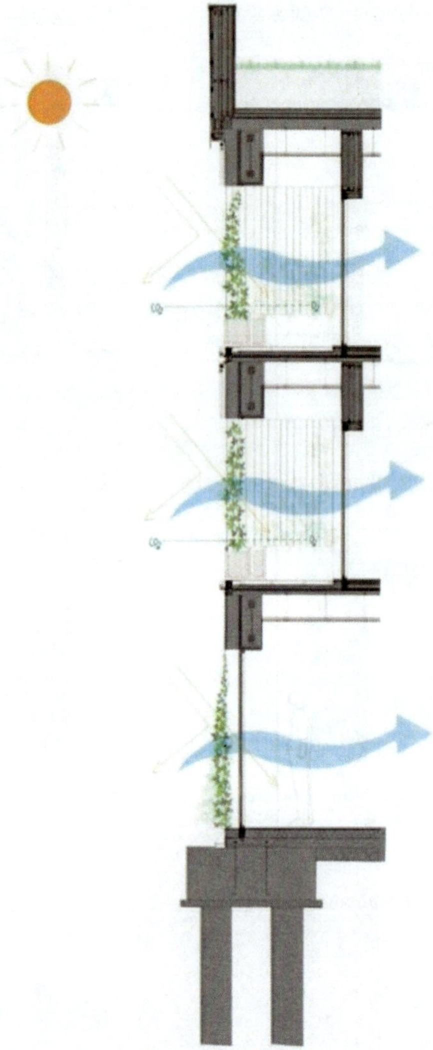

**Fig. 15.5** Green façade energy savings

to the activity of food cultivation. This example was tested by the Sasaki architects the project "Sunqiao Urban Agricultural District Shanghai", in China, 2016 (SASAKI Architects 2016), shown in Fig. 15.7. The architect represents more than a factory for food production. Its masterplan creates a public realm, celebrating agriculture as a key component of urban growth.

**Fig. 15.6** Home farming—the farmhouse concept, PRECHT

**Fig. 15.7** Urban agricultural district, SUNQIAO

## 15.3 The Importance of Water

Water is used a lot in the urban context, for cleaning, sanitation and maintenance of plants or cultivation. Portugal is a country where a lot of rainfall occurs throughout the year, especially in the winter. There has been an increase in flooding, as a result of the climate change, where the rainwater drainage systems are beginning to be insufficient. In addition, there is a large daily consumption of water in homes. A study by the non-profit organisation ANQIP evaluated the average daily consumption in a house in Portugal, Fig. 15.8, which shows that 28% of the water is consumed through flushing (ANQIP).

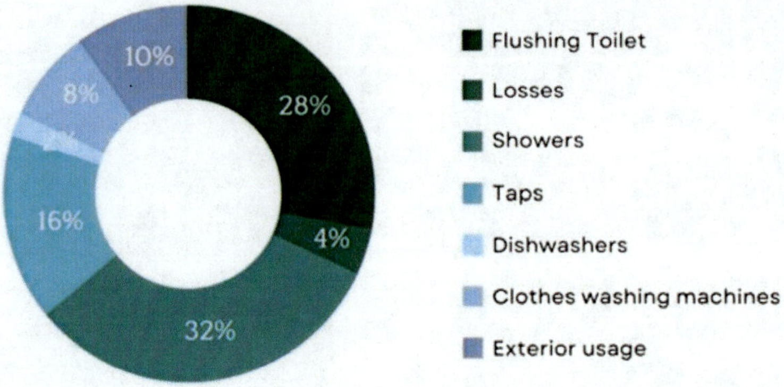


**Fig. 15.8** Average daily water consumption in a house in Portugal, ANQIP

## *15.3.1 Water Collection*

Collecting and absorbing rainwater is not only important for the maintenance of the increased green area, but the whole system can contribute considerably to significantly lessen the impact of extreme weather events (Seak 2021).

A 100 mm drop of precipitation on a 100 $m^2$ area accumulates 10 $m^3$ of water. That's 10 m high in a pond of 1 $m^2$ in area! Without a green area, all this water has to drain somewhere. If in that area there is a 40 $m^2$ garden (divided over several floors), it means that the green area receives 25 cm of water without any amount of water being returned to the sewage system. In other words, a building with 40% of its floor area offset with a green area is able to handle an extreme weather situation (ENVIRONMENT, Corporate-Body 2014). There is another way to collect and store water besides water mirrors on green roofs. Placing a water tank underground, acting as a water reservoir, also helps to control the temperature of a building, Fig. 15.9.

## *15.3.2 Water Reuse*

The process of collecting and storing water, besides helping in extreme weather conditions, can also serve for reuse, for example in sanitary areas. By implementing the reuse of water in flushing toilets we will already be saving 18% of the average daily water consumption. The average monthly water consumption in a family home (4 people) is 10–12 $m^3$, supposing that 18% of the water is used for flushing, then by implementing a water collection tank to reuse the rainwater in the flushing system the family would be sparing around 2–3 $m^3$ of water monthly. Another way of recycling water would be to consider storing the water used in the sinks or showers (called "Greywater") and reuse it after the filtration process in the exterior gardens, through underground tubes, to prevent toxins to reach the air.

**Fig. 15.9** The water collection systems in the demonstration building

## 15.4 Integrated Project Approach

The solutions analysed in this paper were integrated into a demonstration Project. This project is a residential building (7 floors) with an gross area of 2160 $m^2$. Greenhouses implemented throughout the building, made of prefabricated wooden modules, can be placed by the residents in their galleries, serving as home farming, as well as in the public galleries serving the community. Vertical gardens and green facades have also a very positive impact on the energy performance of the building by providing shading and renewable air.

The green roof is another solution that allows a natural cooling of the building's interior, since it does not absorb most of the solar radiation and the building can benefit from the natural evapotranspiration of the green areas. In addition, it contributes to slow the flow of rainwater to the public sewage system, since part of the rainwater will be retained in the soil or in irrigation reservoirs. Therefore it contributes to decrease the potential of urban flooding.

Renewable energy solutions were implemented into the project to minimise the building environmental impact and support the plants growth. The installation of PV panels on the southern façade generates electricity that can be used to power greenhouses lighting and the surplus energy can be sold to the public grid. The PV panels are used as shading devices to protect the facades from direct sun radiation. The PV system is composed of 375 photovoltaic panels, which means that the system will have a total power output of 27 kW. The online PVGis application was used to simulate the performance of the PV system and the results are presented in Table 15.2 and Figs. 15.10, 15.11.

The solution presented for rainwater collection, used for garden irrigation and flushing toilets, contributes not only to the reduction of the soil sealing effect (leading

**Table 15.2** Polysolar panel specification

| Polysolar model | Class | Stabilized performance STC | | | | |
|---|---|---|---|---|---|---|
| | | Transparency | Vmpp (V) | Impp (A) | Voc (V) | Isc (A) |
| PS-CT-72 | 72W | 10% | 87.0 | 0.82 | 116 | 0.88 |

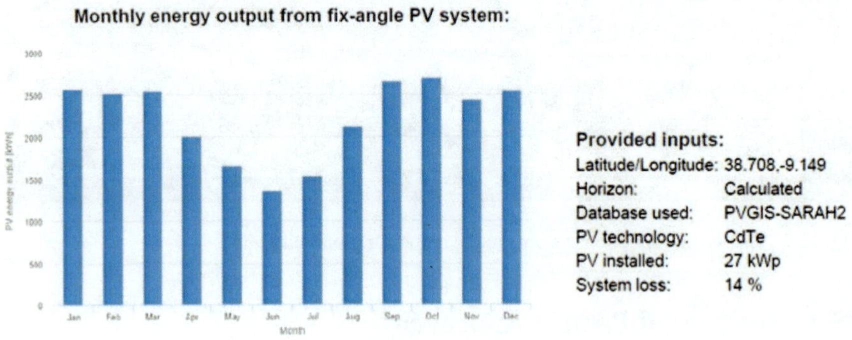


**Fig. 15.10** PVGis online tool results—monthly energy output from fix-angle PV system

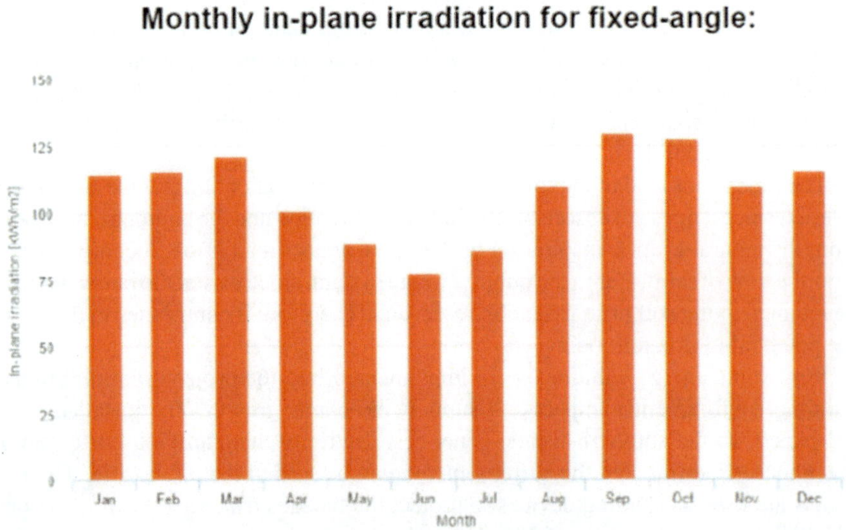


**Fig. 15.11** PVGis online tool results—monthly in plane irradiation, from fix-angle PV system

to floods), but also to a significant reduction in drinking water consumption. The rainwater is collected from the greenhouses roof and directed to a shaft next to the elevators. The water collected through the shafts is conducted to the water storage tank, located on the −1 floor of the building. Using the online tool "Samsamwater.com" it was possible to evaluate how much water is needed per day to irrigate the green area

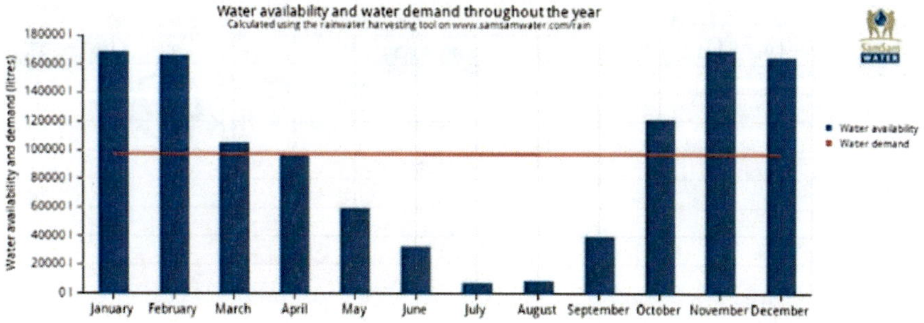


**Fig. 15.12** SamSamwater—online tool—water availability and water demand throughout the year

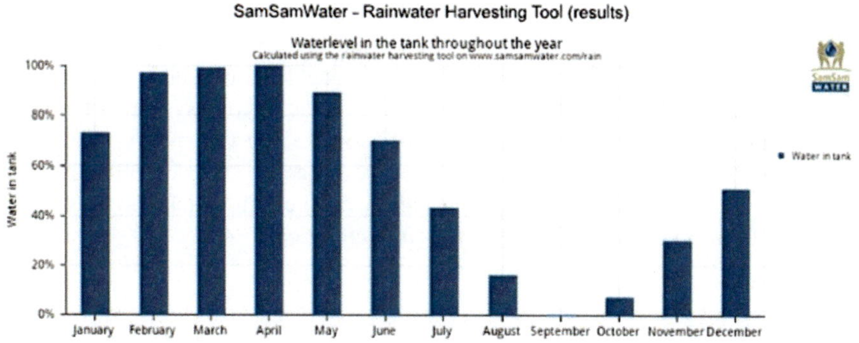


**Fig. 15.13** SamSamwater—online tool—water level in the tank throughout the year

and also calculate how much water the tank can store based on the collection area, Figs. 15.12 and 15.13. 1185 $m^3$ of water are required to maintain the project's green spaces through the year. The total yearly amount of water that can be collected from this roof is 1135 $m^3$, covering 96% of the total demand.

The building also introduces grey water reuse systems. The building therefore has two wastewater drainage pipes where one of them is directly connected to the public wastewater network, and the other to the water filter and storage tank, where after this filtering process, it can be reused. The grey waters account for 50% of the building's water consumption, it is possible to recover around 30% after the filtering system, since the filter needs cleaning to avoid clogging. Assuming an annual water consumption of 4500 $m^3$ of drinking water for the buildings, its is possible to recover around 15% for reuse in the building.

The recovered rain water together with the recycled grey water cover all watering needs of the green spaces for the whole year.

To finalize with another positive environmental feature, this project tries to reduce as much as possible materials that have a high environmental impact such as concrete and brick. The selection of materials is one of the most crucial stages in the design of an architectural project since the carbon footprint, that is emitted into the environment

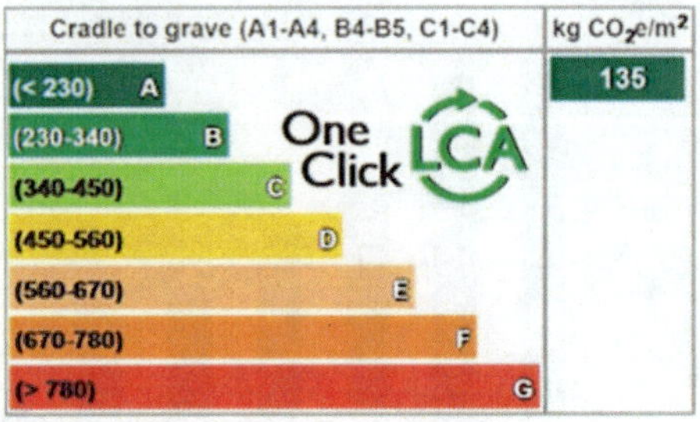


**Fig. 15.14** One Click LCA tool—final building $CO_2$ emitions evaluation

is, in part, related to the material that is chosen. Therefore, the structure of the project is made of steel which can be 100% recycled/reused and finishes in wood which has a positive contribution to the carbon footprint.

The One Click LCA tool takes the materials into account throughout each of the five stages of evaluating the final construction product (product stage, construction stage, use stage, end of life stage and benefits and load beyond system boundary). It was possible to determine the emissions of about 135 kg $CO_2$ per $m^2$, shown in Fig. 15.14. As a comparison the same project was made with a concrete structure and brick walls. The environmental impact of the concrete and brick design is much higher that the steel structure project design, emiting over 249 kg $CO_2$ per square meter.

## 15.5 Discussion

The number of urban gardens has been rising all over the world as a result of a growing demand by city dwellers. This growing demand proves that a solution like the one implemented in this project has a market.

Green-houses throughout the building are one solution to this demand aided by the use of vertical gardens that, without sacrificing practically any living area, have a very positive impact on energy performance by providing shading while contributing to a fresh, healthier environment. The green roof is another solution that takes advantage of an area that is usually wasted, bringing cultivation and leisure activities to the residents of the building. In addition, the introduction of green areas contributes to air purification by increasing biodiversity in the urban space.

The final building covers an implementation area of 2160 $m^2$ (square meter) and offers 3030 $m^2$ of green area, both in horizontal and vertical gardens. This was achieved with an effective reduction of the useful living area of only 30 $m^2$.

The work also uses renewable solutions that support the implemented green area as well as the community. The installation of photovoltaic panels on the southern façade not only generates electrical energy, that can be used to power greenhouses lighting or for injection into the public grid, but also does not occupy floor space.

In any case, it helps to reduce the heating of the facade by solar incidence and to generate clean electricity. This is an investment with guaranteed return.

The building managed to cancel all the waterproofing effects of its implantation area by collecting and storing most of the water in its green area and tanks. On top of that the recovered rain water together with the recycled grey water, cover all watering needs of the green spaces for the whole year. This solution presented for rainwater collection, used for garden irrigation and flushing, contributes not only to the reduction of the soil sealing effect, but also to a significant reduction in drinking water consumption.

The use of prefabrication for some parts of the building and its structure also contributes to faster construction, reducing the amount of work to be done on site and reducing the emission of polluting gases into the environment. The building offers a quality A environmental impact when it comes to its construction and lifetime evaluation, using the One Click LCA calculator tool.

All these features help mitigate the negative impact of architecture on the urban environment.

## 15.6 Conclusion

The introduction of these green elements in urban space is therefore responsible for mitigating the heat island effect and contribute to the fixation of pollutants as well as reduce noise levels, slow the flow of rainwater, prevent the risk of flooding, and decrease energy consumption through better thermal insulation of buildings. On top of that, the green elements can be adjusted accordingly to the residents' wishes to ensure energy savings (shading and cooling) and even provide them with food, by making the food growing process a public facility. In addition, it will promote the societal interactions between neighbours.

The work presented here proves that it is possible to restore a significant area of green in a residential building in an urban context, even exceeding its footprint. The work also shows that the habitable area to be sacrificed is considerably smaller than one might initially assume, making it a positive solution. In this project, a circular balance was achieved between habitability and nature, without compromising the construction costs or profitability of the building. It is therefore possible to build without taking away green areas, with positive results for people and the environment.

## References

AMORIM, Ana Francisca Fernandes Ferreira (2015) Vegetation in the exterior surroundings of buildings: impacts, constraints and eco-efficient intervention strategies. University of Minho, School of Architecture (in Portuguese)

ANQIP—Construction of a rainwater utilization system. New building for the order of architects (northern region). https://adaptis.uc.pt/articles/43

Bennet CC (2013) Vertical gardens a green solution for urban setting. The Times of India, Ltd. Retrieved 20 Feb 2013

Blank P (2008) The vertical garden: from nature to the city. W. W. Norton, New York

ENVIRONMENT, Corporate-Body (2014) E.-G. for.—Orientações sobre as melhores práticas para limitar, atenuar ou compensar a impermeabilização dos solos. Office of the EU

European Conference (1994) Charter of European cities & towns towards sustainability, Denmark

ITECONS (2015) Adaptation measures: Green roofs (in Portuguese), Portugal. https://adaptis.uc.pt/articles/21. Accessed Oct 2015

PRECHT (2019) The farmhouse concept

RAPOSO, Fausto Miguel Ferreira (2013) Green roofing best practices manual (Case study analysis). Construction and rehabilitation (in Portuguese)

SASAKI Architects (2016) Sunqiao urban agricultural district Shanghai. China

Seak TH (2021) Tangu architecture Sdn Bhd. Permeability housed/Tangu architecture. Kuala Lumpur

Simões RC (2020) "Growing in the building" Sustainability of urban gardens integrated into green roofs and facades. University of Minho, School of Architecture (in Portuguese)

Stiles W, Wootton-Beard P (2017) Vertical farming: a new future for food production? System Industrial measurement, S1. IBERS, Aberystwyth University, pp 421–423 (in Chinese)

The International Greenroof & Greenwell Projects Database (2013) Select 'green wall' as type and 'living wall' under 'green roof type'. Greenroofs.com. Retrieved 17 Oct 2013

Touliatos D, Dodd IC, Martin M (2016) Vertical farming increases lettuce yield per unit area compared to conventional horizontal hydroponics. The Lancaster Environment Centre, Lancaster University, Lancaster, UK

# Chapter 16
# Study on the Current Spatial Landscape of the Moat in the Western Section of Jinan Mingfu City

**Wei Wu and Kexin Ding**

**Abstract** Water is the most characteristic urban natural landscape of Jinan. Jinan's urban history and local customs are all most directly and intimately related to the spring. With the change of history, the pattern of the moat surrounding the city in Mingfu City has gradually evolved into a moat embedded inside the city. In the process of this transformation, the moat was gradually integrated with the road and further separated from the residents' lives, gradually detaching itself from people's daily lives. As an important connection carrier of Jinan's spring system, the moat carries profound historical and humanistic values and possesses rich historical and cultural connotations and unique spatial characteristics.

**Keywords** Spatial landscape · The moat · Jinan Mingfu City

## 16.1 Overview of Jinan Moat

Jinan moat along with the development of Jinan, precipitated a deep historical and cultural elements, and Jinan's modern development is closely related to the special space that nurtured the development of Jinan city function. The tangible and intangible traces of its history reveal the cultural identity of the city. Figure 16.1 shows the spatial pattern of the moat in the Mingfu City area of Jinan.

Jinan's moat has a long history. 600 years ago, there was already a fairly wide moat around the periphery of Jinan's city walls. At that time, the moat, "a week around the city, the pool is five feet wide, the water is three feet deep, between the moat and the city walls. Remaining moat about a few feet wide." As the water source of the moat Baotu, Black Tiger Springs, two large spring group gushing all year round, plus the moat of other water sources, from the south of the city Thousand Buddha Mountain, Four Mile Hill, Swallow Wing Hill and other valleys flow to the

W. Wu (✉) · K. Ding
School of Jinan University, Jinan, China
e-mail: 3026576474@qq.com



P. Mendonça et al. (eds.), *Proceedings of 2023 International Conference on Green Building*, Environmental Science and Engineering,
https://doi.org/10.1007/978-3-031-43478-5_17

**Fig. 16.1** Spatial structure of the Jinan moat

city of water, through the culvert injected into the moat. The Jinan moat has been well-springing and stable throughout its history. It not only plays a role in protecting the city militarily, but also adds Jiangnan charm to Jinan, the city of springs, with its water and scenery along the river. But with the construction of the city, the scene of intimate interaction between the people along the moat and the river no longer exists. Figure 16.2 shows the scene of the western section of the Jinan moat in 1929 (Jinyu and Xin 2018).

Jinan Moat The Jinan Moat is 6.9 km long, with an area of 26.3 hectares. The width of the river varies from 10–30 m, with a water surface area of 8.4 hectares, and green areas on both sides of the river from 10–59 m, with a green area of 12.5 hectares.

West section of the moat is Jinan Mingfu City moat among the surrounding landscape and function of the richest section. The western section of the moat north of Daming Lake, north–south direct distribution, through residential areas, commercial areas, historical and cultural districts, city parks, south to the Baotu and Quancheng Square. The wide range of radiation, covering a number of functions unique.

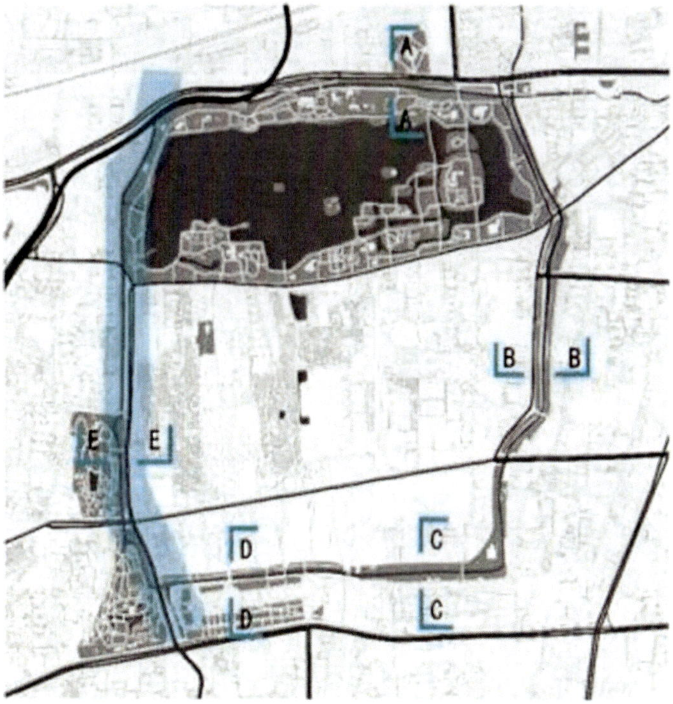


Fig. 16.2 Original view of the moat

## 16.2 The Spatial Situation of the Jinan Moat

### 16.2.1 Spring Space

The moat is the richest natural area in Jinan, with complex ecological structure and function, and a good moat environment is a key link to improve the ecological quality of the city. The moat is not only a moat but also a "cultural river", as it connects all the major springs in Jinan, which have been the source of famous quotes from literati and scholars throughout the ages. The springs have a variety of forms and spatial relationships, creating a rich spatial relationship. Table 16.1 presents the different spatial relationships between buildings, streets, and springs, which are diverse and charming (Wenzhi and Xiaoming 2022).

**Table 16.1** Spring space type

| Spring form | Location Relationship | Live Picture | Relationship with the street | Relationship with public space |
|---|---|---|---|---|
| Fountainhead |  |  | Using the building façade as an interface | The spring is located in one corner of the square,with the building facade as the interface on both sides |
| Spring Road |  |  | Use the building facade as the interface on one or both sides | The node is enlarged to become a public space on the street where people are concentrated |
| |  |  | Independent of the building and parallel to the street | The semi-private space serves as a transition between the street and the occupants |
| Spring Pond |  |  | Independent of the building, surrounded by roads | The combination of building and courtyard together enclose the spring pond, giving the space both a sense of enclosure |
| | 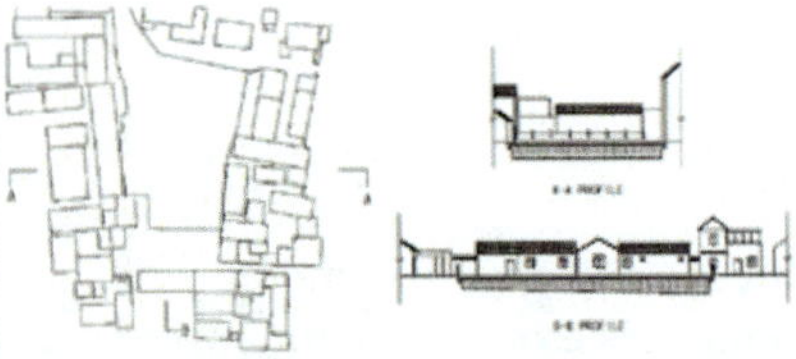 | | | |

## 16.2.2 *River Space*

The river moat space echoes the surrounding environment, buildings and structures, forming a rich variety of spatial types. At present, the historical function of Jinan moat is disappearing, while its new function definition has not yet been formed, and its role in the city has not yet been fully played, so there are still some shortcomings.

**Vertical riverbank.** The vast majority of the moat is a stone masonry vertical embankment that takes away from the vitality of the urban public space of the moat. Figure 16.3 shows the cross-section of the moat and its surrounding areas in Fig. 16.1. The tall, vertical, hard stone embankments make it difficult to create

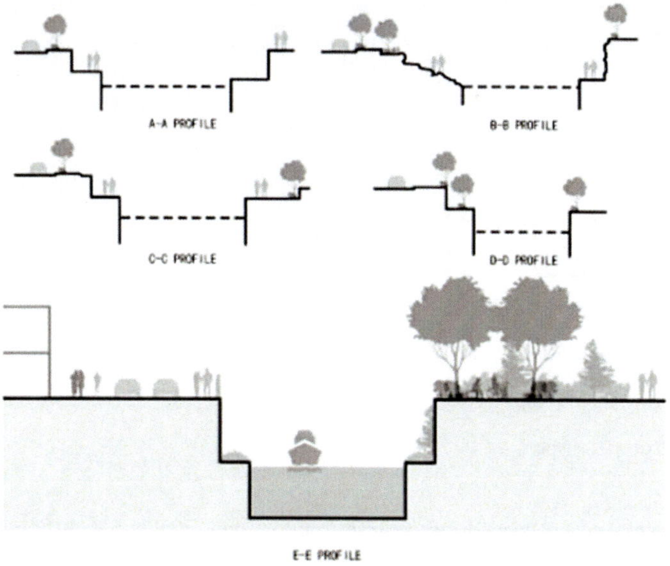


**Fig. 16.3** Moat cross-section

a pleasant recreational space because the area loses its intimacy, causing visitors to rarely stay and cutting off the life of the residents along the river and the moat. The hard vertical embankment is not conducive to the healthy development of the moat ecology, and seriously affects the continuity of living conditions along the riverbank and the formation of the river landscape image.

**Vertical riverbank.** Secondly, the Jinan moat area, in the center of the city, is one of the most densely trafficked areas, with multiple sections of roads and bridges crossing the river, but without taking into account the needs under the bridges, tourist routes often end up at the end, creating many "break points". The viaducts in the northwest part of the moat completely cover the river, almost becoming an underground river. The busy urban traffic has severely damaged the peaceful and relaxing moat landscape.

### 16.2.3 Riverfront Space

The riverfront space is variable in form, and there is a height difference between the moat and the city road, presenting a multi-level spatial sightline. Figure 16.4 shows three spatial forms of the western section of the moat. Compared with the flat landscape, the change of viewpoint and field of vision in vertical direction is more significant, which makes the circumcity park present a thousand inward-looking visual space in the topographic structure spatial scale, with close internal visual

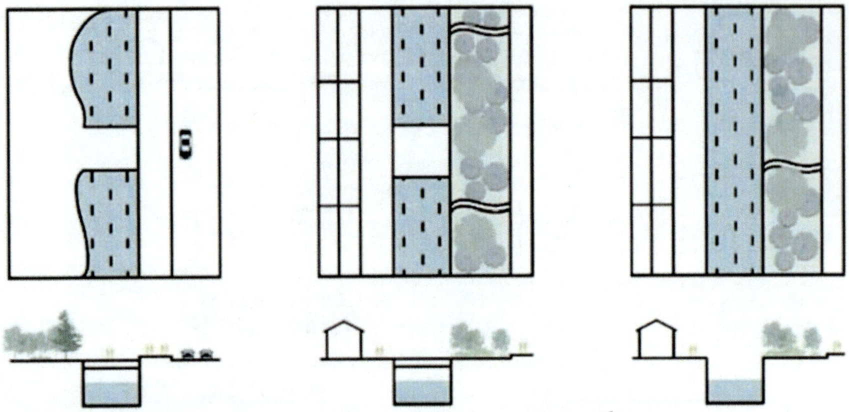

Fig. 16.4 Moat space type

connection, closed space and concentrated sight lines, making it difficult to peep and perceive the landscape of the moat and the pattern of the old city of Jinan from outside.

The visual accessibility perpendicular to the water flow is an important design consideration. For citizens in the inner city space, this view is more directly connected to the waterfront open space, and through landscape guidance and attraction, it awakens the desire of citizens to go to the waterfront open ssspace. The visual accessibility perpendicular to the flow of water is closely related to the permeability of the sight lines and view corridors (Xiangrong 2008).

## 16.3 Conclusion

The moat of Jinan is the most vibrant and vital space of the city, and is also a witness of the continuation of the city's cultural lineage. By shaping a beautiful urban waterfront environment, preserving the traditional memory of Jinan's historical areas, and renewing the moat to build a perfect modern function, it is necessary to reshape the characteristics of the spring city and stimulate the economic vitality of the region and the city.

In the redesign planning of the moat, water-friendly features should be highlighted to make it a multi-element river route combining tourism landscape and leisure landscape. The original factories along the riverbank can be transformed into creative industrial parks, and through selective preservation, renovation, demolition and new construction, the moat can be made into an art road of Jinan's history and vernacular scenery. Facilities such as galleries, artists' studios, bars, restaurants, art stores, bookstores, outdoor sculpture plazas and gardens can be set up along the riverbank to expand the local cultural atmosphere of the moat, open a coastal articulation pier, combine with river yachts to introduce and understand Jinan's history and culture to visitors, and add water play areas and water-friendly activity areas for visitors (Tao and Zheng 2022).

## References

Liu Jinyu, Jiang Xin (2018) Research on the visual reachability of the waterfront space of the Jinan moat [C]. In: China Urban Planning Society, Hangzhou Municipal People's Government. Sharing and Quality—Proceedings of the 2018 China Urban Planning Annual Conference (13 Landscape Environmental Planning). China Architecture Industry Press, s8

Liu Tao, Fang Zheng (2022) A brief discussion on the planning and design project for the protection of the inner and outer rivers of Jingzhou Ancient City [J]. Engineering Construction and Design, No. 484 (14): 45–47. https://doi.org/10.13616/j.cnki.gcjsysj.2022.07.214

Xiangrong W (2008) Lin Qing Water bank revival of "Spring City"—landscape planning along the Daming Lake and moat in Jinan [J]. Chinese Garden 24(12):33–38

Xu Wenzhi, Gao Xiaoming (2022) A study on the creation of urban public space places based on behavioral patterns: taking the waterfront Space of Jinan City Protector as an Example [J]. Urban Architectural Space, 29(12): 124-126

# Chapter 17
# Improved Analysis System for Determining the Effectiveness of Natural Smoke and Heat Exhaust Ventilator

**Hang Yin and Longfei Tan**

**Abstract** In the design field of fire engineering, the effectiveness determination of each natural smoke and heat exhaust ventilator should be as accurate as possible to prevent the insufficient capability or overcapacity of the natural smoke and heat exhaust ventilation system. Aiming at the limitations of current mainstream analysis mechanisms, an improved analysis system is built from the perspective of better considering various influence factors and closely combining with the design conditions of system. The analysis mechanisms to determine the coefficient of discharge of natural smoke and heat exhaust ventilator under the conditions without and with side wind are proposed in a more scientific and reasonable way, which is helpful in establishing the effectiveness database of all typical natural smoke and heat exhaust ventilator in the future.

**Keywords** Analysis system · Effectiveness · Natural smoke and heat exhaust ventilator · Coefficient of discharge · Design condition

## 17.1 Introduction

For the design of natural smoke and heat exhaust ventilation system (NSHEVS), it is critical to precisely determine the effectiveness of each individual natural smoke and heat exhaust ventilator (NSHEV). The overestimate or underestimate of the effectiveness of any NSHEV may lead to the insufficient capability or overcapacity of NSHEVS. Up to now, most studies focused on the influence factors of natural smoke

H. Yin (✉)
Sichuan Fire Research Institute of Ministry of Emergency Management, Chengdu, China
e-mail: bank2002yinhang@126.com

L. Tan
School of Emergency Management, Xihua University, Chengdu, China
e-mail: longfei_tan@163.com



P. Mendonça et al. (eds.), *Proceedings of 2023 International Conference on Green Building*, Environmental Science and Engineering,
https://doi.org/10.1007/978-3-031-43478-5_18

and heat exhaust ventilation on the system or facility level, which involve the critical wind speed (Chen et al. 2009; Yi et al. 2013), adjacent wall (Li et al. 2014; Chow and Li 2018), roof obstacles (Węgrzyński et al. 2018, 2019) and comprehensive factors (Król and Król 2017). Few studies discussed the test method for the effectiveness of NSHEV within the framework of current standards. However, there were relatively less studies regarding the analysis mechanism for determining the effectiveness of NSHEV. In this study, an improved analysis system is built from the perspective of better considering various influence factors and closely combining with the design conditions of NSHEVS.

## 17.2 Current Analysis Systems and Corresponding Limitations

### *17.2.1 Analysis Mechanisms of Current Standards*

The effectiveness of individual NSHEV is determined by calculating the smoke exhaust effective area for the mainstream analysis systems, using the following equation:

$$A_e = A_v \bullet C_v \tag{17.1}$$

where $A_e[\mathrm{m}^2]$, $A_v[\mathrm{m}^2]$ and $C_v[\text{-}]$ are the smoke exhaust effective area, the opening area and the coefficient of discharge, respectively.

The key point in the accurate determination of aerodynamic free area is whether or not all the influence factors and design conditions are considered sufficiently during the test and analysis of the coefficient of discharge of NSHEV, which is determined using either of the following equivalent equations:

$$C_v = \frac{V}{A_v}\sqrt{\frac{\rho}{2\Delta p}} \tag{17.2}$$

$$C_v = \frac{m}{A_v\sqrt{2\rho\Delta p}} \tag{17.3}$$

where $V[\mathrm{m}^3/\mathrm{s}]$, $\rho[\mathrm{kg/m}^3]$, $\Delta p[\mathrm{Pa}]$ and $m[\mathrm{kg/s}]$ are the volume flow rate of gas medium, the density of gas medium, the pressure drop across the NSHEV and the mass flow rate of gas medium, respectively.

For USA, according to the standard of NFPA 204, the coefficient of discharge of NSHEV can be obtained from the manufacturer directly or be taken from the data of some specific NSHEV types provided by the standard.

For Australia, according to the standard of AS 2428.5, the coefficient of discharge of NSHEV should be determined by experimental testing. The final value of coefficient of discharge is determined by averaging at least six sets of data corresponding to different flow rates of air.

For Europe, according to the standard of EN 12,101–2, the combination of simple and experimental assessment is used. In the procedure of simple assessment, for some specific roof mounted NSHEV, the reference values of coefficient of discharge are suggested as a fixed value of 0.4; for two bottom-hung wall mounted NSHEVs, outward and inward opening, the discharge coefficients for various opening angles are directly provided by the standard. In the procedure of experimental assessment, the experimental testing should be conducted for the conditions both without and with side wind. In terms of the test conditions without side wind, similarly to Australian method, the calculated coefficient of discharge should be the mean value of at least six sets of data. But the pressure drop across the NSHEV for each flow rate of air shall be within the range of 3 Pa ~12 Pa. For the test scenarios with side wind, at least six sets of tests shall be conducted to each wind direction under the premise of maintaining the pressure drop across the NSHEV between 0.005 ~0.15 times dynamic pressure of side wind. Then the calculated coefficient of discharge at certain wind direction can be determined from the regression line as the value corresponding to the pressure drop across the NSHEV equal to 0.082 times dynamic pressure of side wind. The finial determined coefficient of discharge is the minimum value for all conditions with and without side wind.

### *17.2.2 Limitations of Current Analysis Systems*

For the analysis systems of both Australian and European standards, all the parameters participated in the calculation belong to the air under normal temperature driven by mechanical power of fan, the characteristic of which is quite different from that of the high-temperature smoke driven by buoyancy in fire scene.

Moreover, neither of above-mentioned analysis systems combines with the design conditions of NSHEVS. For the analysis mechanism of the test condition without side wind, the mean value of the calculation results under different flow rates of air is regarded as the final value of discharge coefficient. However, the coefficient of discharge of NSHEV is insensitive to the mass flow rates of air under normal temperature. The discrepancy of different design conditions caused by the differences in smoke densities is not reflected during the analysis procedure. For the analysis mechanism of the test condition with side wind, only the condition with a fixed wind velocity equal to or greater than 10 m/s is considered, the effects of different side wind velocities on the coefficient of discharge of NSHEV are ignored. In addition, the fitted value corresponding to the pressure drop across the NSHEV equal to 0.082 times dynamic pressure of side wind is taken as the final determined value of discharge coefficient. Nevertheless, the ratio of pressure drop across the NSHEV to the dynamic pressure of side wind is not fixed under the interaction between indoor fire environment and outdoor wind environment for different design conditions of NSHEVS.

## 17.3 Improved Analysis System Based on Design Conditions of NSHEVS

For the limitations of current analysis systems, an improved analysis system is built as follows. It is noted that a test apparatus, which has all the functions necessary to supply the fire source of different heat release rates and side wind of different velocities and directions, is essential for acquiring required parameters participated in the corresponding calculations. The form of the test apparatus can be, but not be limited to, the following structural style, as shown in Fig. 17.1.

### *17.3.1 Analysis System for the Condition Without Side Wind*

First, the smoke density of smoke layer can be expressed as the following equation based on the ideal gas law:

$$\rho_s = \frac{P}{RT_s} \tag{17.4}$$

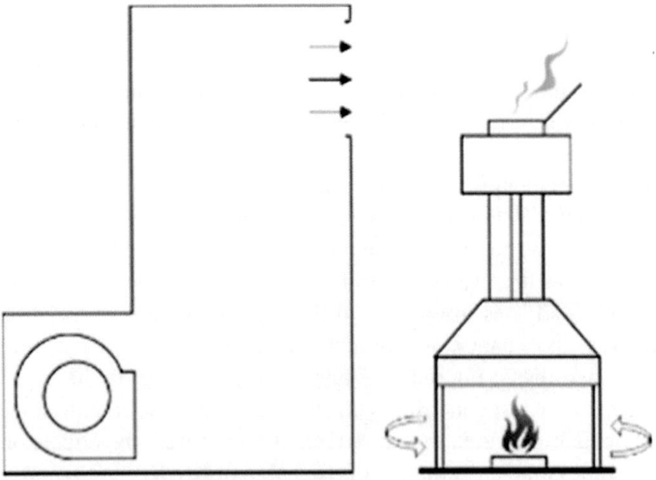

**Fig. 17.1** The schematic diagram of the test apparatus

where $\rho_s$[kg/m$^3$], $P$[Pa], $R$[-] and $T_s$[K] are the smoke density of smoke layer, the pressure of smoke layer, the gas constant of smoke and the smoke temperature of smoke layer, respectively. The $P$ and $R$ can be approximately replaced by the atmospheric pressure and gas constant of air, respectively, for the engineering application.

Combining Eq. (17.3) with Eq. (17.4), the coefficient of discharge of NSHEV can be determined using the following equation:

$$C_v = \frac{m}{A_v}\sqrt{\frac{T_s}{706\Delta p}} \quad (17.5)$$

Measure the $m$, $T_s$ and $\Delta p$ over a period under the condition of different heat release rates of fire source, then calculate each $C_v$ by substituting the time averages of above-mentioned parameters into Eq. (17.5). The function relationships of $C_v = f(\Delta p)$ and $Q = f(\Delta p)$ can be determined by curve fitting, as shown in Fig. 17.2 and Fig. 17.3.

According to Eq. (17.3), the pressure drop between smoke layer and outdoor environment under the design condition of NSHEVS can be calculated as follows:

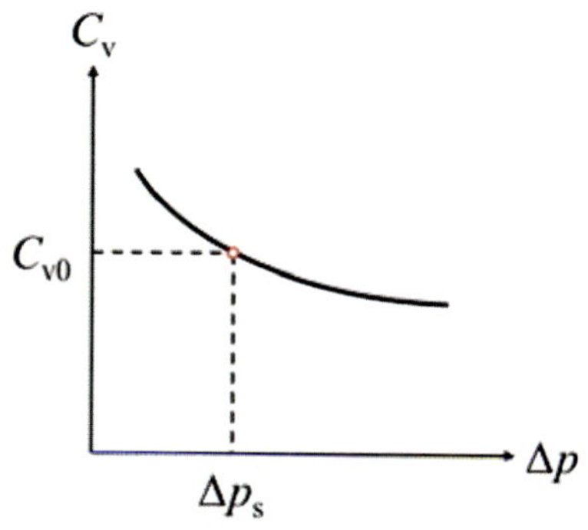


**Fig. 17.2** The function relationship between $C_v$ and $\Delta p$ and the determination of $C_{v0}$

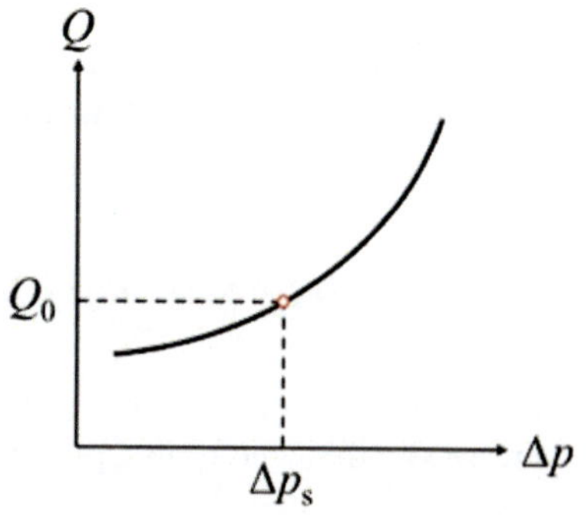


**Fig. 17.3** The function relationship between $Q$ and $\Delta p$ and the determination of $Q_0$

$$\Delta p_s = \frac{M^2}{2\rho_s A_{e,total}{}^2} \tag{17.6}$$

where $\Delta p_s$[Pa] is the pressure drop between smoke layer and outdoor environment, $M$[kg/s] and $A_{e,total}$[m$^2$] are the mass flow rate entering the smoke layer and the minimum required aerodynamic free area of NSHEVS, respectively, which can be determined using the following equations based on the design condition of NSHEVS:

$$z_e = 0.166 Q_c^{2/5} \tag{17.7}$$

$$M = \begin{cases} 0.071 Q_c^{1/3} z^{5/3} + 0.0018 Q_c, z > z_e \\ 0.032 Q_c^{3/5} z, z \le z_e \end{cases} \tag{17.8}$$

$$T_s = T_0 + \frac{K_s Q_c}{M C_p} \tag{17.9}$$

$$A_{e,total} = \frac{M}{\rho_0}\left[\frac{T_s{}^2 + \left(\sum_{i=1}^{n} A_{v,i} C_{v,i} / \sum_{j=1}^{k} A_{in,j} C_{in,j}\right)^2 T_s T_0}{2gd(T_s - T_0)T_0}\right]^{1/2} \tag{17.10}$$

where $z_e$[m], $Q_c$[kW], $z$[m], $T_0$[K], $K_s$[-], $C_p$[kJ/(kg·K)], $\rho_0$[kg/m$^3$], n[-], $A_{v,i}$[m$^2$], $C_{v,i}$[-], $k$[-], $A_{in,j}$[m$^2$], $C_{in,j}$[-], $g$[m/s$^2$] and $d$[m] are the flame height, the convective portion of the heat release rate of the fire source, the distance above the base of the fire source, the ambient temperature, the fraction of convective heat release contained in smoke layer, the specific heat of smoke, the ambient density, the number of NSHEV in a NSHEVS, the geometric area of the $i$'th NSHEV, the coefficient of discharge of the $i$'th NSHEV, the amount of make-up air inlet in a NSHEVS, the opening area of the $j$'th make-up air inlet, the coefficient of discharge of the $j$'th make-up air inlet, the acceleration due to gravity and the smoke layer depth.

Then the coefficient of discharge of NSHEV under the condition without side wind, $C_{v0}$, can be determined by substituting $\Delta p_s$ into the function of $C_v = f(\Delta p)$. The required heat release rate of fire source to fulfil the pressure drop across the NSHEV corresponding to the condition without side wind, $Q_0$, can be determined by substituting $\Delta p_s$ into the function of $Q = f(\Delta p)$.

### *17.3.2 Analysis System for the Condition with Side Wind*

The side wind velocity at the height of NSHEV from the outdoor ground can be expressed by the wind profile power law:

$$v_H = v_{10}\left(\frac{H}{10}\right)^{\alpha} \tag{17.11}$$

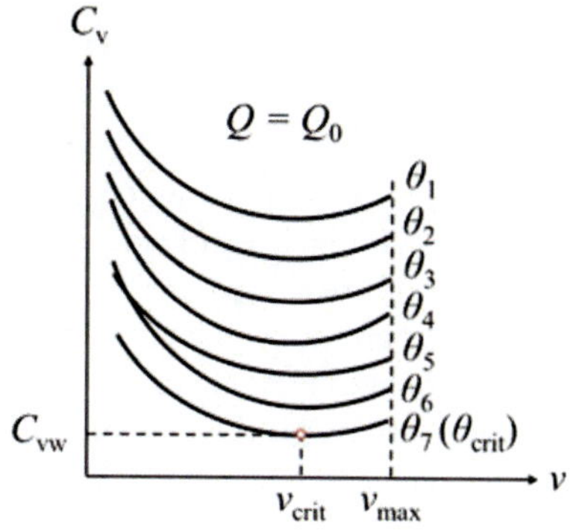


**Fig. 17.4** The function relationship between $C_v$ and $v$ and the determination of $C_{vw}$

where $v_H$[m/s], $v_{10}$[m/s] and $\alpha$[-] are the velocity of side wind at the height of $H$ and 10 m from the outdoor ground, and the exponential parameter, respectively. The $v_{10}$ and $\alpha$ can be determined by the meteorological data and terrain category, respectively.

Assign the value of Q0 to the heat release rate of fire source and carry out the experiments at various wind directions. For each wind angle, measure the m, Ts and Δp over a period under the condition of different velocities of side wind. The maximum velocity vmax shall be no less than the calculated wind velocity at the height of NSHEV from the outdoor ground. Calculate each Cv by substituting the time averages of above-mentioned parameters into Eq. (17.5). The function relationships of $C_v = f(v)$ corresponding to each wind angle can be determined by curve fitting, as shown in Fig. 17.4.

Calculate the minimal value of each function, the minimum coefficient of discharge of NSHEV under the condition with side wind, Cvw, can be determined corresponding to the vcrit and θcrit, which are the most disadvantageous velocity and angle of the side wind for the natural smoke and heat exhaust ventilation of NSHEV.

### *17.3.3 Determination of the Minimum Effectiveness of NSHEV*

The minimum effectiveness of NSHEV under the design condition of NSHEVS can be determined by substituting the smaller value between Cv0 and Cvw into Eq. (17.1).

## 17.4 Conclusion

Any inaccurate or incorrect assessment on the effectiveness of NSHEV should be avoided for the engineering design of NSHEVS, especially for the overestimate of effectiveness which will lead to the inadequate capability of the NSHEVS and the insecurity of fire escape.

For the current mainstream analysis mechanisms, all the parameters participated in the calculation belong to the air under normal temperature driven by mechanical power of fan, rather than the high-temperature smoke driven by buoyancy. In addition, the testing conditions of NSHEV are independent of the design conditions of NSHEVS, which may cause the significant difference between the analysis results and true values of the effectiveness of NSHEV.

In this study, an improved analysis system is built in a more scientific and reasonable way. The pressure drop between smoke layer and outdoor environment under the design condition of NSHEVS is regarded as the key parameter for linking up the design condition with testing condition. The coefficient of discharge of NSHEV under the condition without side wind and the minimum coefficient of discharge of NSHEV under the most unfavorable condition without side wind are determined, respectively, based on the design condition of NSHEVS.

It can be expected that the effectiveness database of all typical NSHEVs in all situations will be established according to the proposed new analysis system in the future.

**Acknowledgements** This work was supported by Research Program of Fire and Rescue Department of Ministry of Emergency Management (No.2018XFGG18) and Central Public-interest Scientific Institution Basal Research Fund (NO.20218805Z).

## References

Chen H, Liu N, Chow W (2009) Wind effects on smoke motion and temperature of ventilation-controlled fire in a two-vent compartment. Build Environ 44:2521–2526

Chow WK, Li J (2018) Wind action on natural smoke exhaust in atria. Journal of Computational Science 28:140–147

Król M, Król A (2017) Multi-criteria numerical analysis of factors influencing the efficiency of natural smoke venting of atria. J Wind Eng Ind Aerodyn 170:149–161

Li H, Fan C, Wang J (2014) Effects of wind and adjacent high rise on natural smoke extraction in an atrium with pitched roof. APCBEE Proc 9:296–301

Węgrzyński W, Krajewski G, Kimbar G (2018) A concept of external aerodynamic elements in improving the performance of natural smoke ventilation in wind conditions. AIP Conf Proc 1922:110006

Węgrzyński W et al (2019) The influence of roof obstacles on the performance of natural smoke ventilators in wind conditions. J Wind Eng Ind Aerodyn 189:266–275

Yi L et al (2013) Study on effect of wind on natural smoke exhaust of enclosure fire with a two-layer zone model. J Wind Eng Ind Aerodyn 119:28–38